THE GOOD CAT PARENT'S GUIDE TO FELINE BEHAVIOR MODIFICATION

Cats are cuddly and adorable, but they are often misunderstood. Sadly, many cats are relinquished to shelters or rehomed due to normal behaviors that are incorrectly treated or mishandled. In this book, Elite Fear-Free and Low-Stress Handling Certified author Alana Linsay Stevenson empowers cat parents and teaches them how to address and modify challenging feline behavior.

You will begin by learning basic kitten care and feline developmental stages; how cats differ behaviorally from group animals, such as dogs and people; feline body language; and how cats handle stress. Alana provides concise instruction on how to gently handle cats: how to pick up and carry them, acclimate them to carriers, the use of towels, alternatives to scruffing, and how our body language affects cats. Packed with photographs for visual reference, this book offers clear guidelines and easily implementable strategies for resolving feline behavioral problems, such as:

- failure to use the litter box
- play aggression
- petting aggression
- inter-cat aggression
- furniture scratching
- jumping on counters
- obsessing about food
- night wailing
- fear of people
- aggression to strangers.

The content is organized by topic for easy access to information, as you need it.

The Good Cat Parent's Guide to Feline Behavior Modification is for anyone who likes cats and wants to learn more about them. Whether you are a veterinary professional, a volunteer or shelter worker who regularly handles stressed cats, or a cat parent who simply wants to understand your cat, you will find helpful and useful information at your fingertips to give cats a better quality of life. No cat parent should be without this book!

THE GOOD CAT PARENT'S GUIDE TO FELINE BEHAVIOR MODIFICATION

Alana Linsay Stevenson

CRC Press is an imprint of the
Taylor & Francis Group, an **informa** business

First edition published 2024
by CRC Press
6000 Broken Sound Parkway NW, Suite 300, Boca Raton, FL 33487-2742

and by CRC Press
4 Park Square, Milton Park, Abingdon, Oxon, OX14 4RN

CRC Press is an imprint of Taylor & Francis Group, LLC

Library of Congress Cataloging-in-Publication Data
Names: Stevenson, Alana, author.
Title: The good cat parent's guide to feline behavior modification / Alana Linsay Stevenson.
Description: First edition. | Boca Raton : CRC Press, 2024. | Includes bibliographical references and index. | Summary: "In this book, Elite Fear-Free and Low-Stress Handling Certified author Alana Stevenson aims to empower cat parents by demonstrating how to address and modify challenging cat behaviors. You will begin by learning basic kitten care and feline developmental stages, how cats differ behaviorally from group animals such as dogs and people, feline body language and how cats handle stress. The book then provides clear guidelines for resolving feline behavioral problems such as failure to use the litter box, play aggression, petting aggression, inter-cat aggression, furniture scratching, jumping on counters, obsessing about food, night wailing, fear of people, and aggression to strangers"— Provided by publisher.
Identifiers: LCCN 2023006696 (print) | LCCN 2023006697 (ebook) | ISBN 9781032398761 (paperback) | ISBN 9781032398778 (hardback) | ISBN 9781003351801 (ebook)
Subjects: LCSH: Cats—Behavior. | Cats—Behavior therapy.
Classification: LCC SF446.5 .S74 2024 (print) | LCC SF446.5 (ebook) | DDC 636.8/083–dc23/eng/20230428
LC record available at https://lccn.loc.gov/2023006696
LC ebook record available at https://lccn.loc.gov/2023006697

ISBN: 9781032398778 (hbk)
ISBN: 9781032398761 (pbk)
ISBN: 9781003351801 (ebk)

DOI: 10.1201/9781003351801

Typeset in Bembo
by codeMantra

This book is dedicated to cats. All the cats who are misunderstood and suffering, all the cats who are homeless, and all the cats who are well-loved and cared for.

CONTENTS

PREFACE

Cats are loved, but misunderstood. People consider them untrainable. Many cats are brought to shelters or abandoned due to normal feline behavior that people misunderstand or handle inappropriately, or because of behavior problems that go untreated or that are treated incorrectly. Yet we have so much influence over our cat's behavior. Our responses and *our* behavior can change a cat's behavior, for better or for worse. When using misinformed or punitory approaches, we can ruin the relationship we have with our cats and create more anxiety and stress. We can cause our cats to avoid the litter box and make aggressive play behavior and other forms of feline aggression worse, not better. We can even inadvertently make our cats hate each other. Or, we can encourage our cats to get along with each other, to use the litter box regularly, and to be calmer and more affectionate with us. We hold so much potential to help, or hinder, our cats' well-being and behavior.

Based on my own experience rehabilitating and trapping cats, and living with them since childhood, as well as an additional 20 years' experience professionally going into clients' homes and helping them to resolve feline behavior problems, I know what the common problems and complaints are, and where and how people get discouraged. I am aware of the popular techniques frequently found online and regularly recommended; techniques that do far more damage to cats than help them.

Understanding cats makes us appreciate them more. They are unique individuals who should be respected simply because 'they are.'

Hopefully, this book will help you to resolve any behavioral issues you may be having with a cat and to better understand them.

This book is designed so you can skip to chapters or sub-sections as you need them. However, I strongly recommend reading the introduction or first few pages of each chapter before you skip to a section within it. This will provide you with a better foundation and understanding of why certain recommendations are made. For the sake of this book and clarity, 'punishment' means something the cat dislikes or considers aversive. This is the standard use of the word that the public generally accepts and is familiar with. Likewise, a 'stressor' or 'trigger' is something that causes stress, which is how most people interpret it. If you have a specific interest in more detailed behavioral terminology and definitions, see Glossary/Behavioral Terminology in the back of this book.

Any unattributed photos are my own.

ABOUT THE AUTHOR

Alana Linsay Stevenson, M.S. is an animal behaviorist and dog trainer. She helps her clients resolve dog and cat behavioral problems, including fearful and aggressive behaviors, in a humane, science-based way.

Alana has a Master of Science in Biology Education and a Bachelor's degree in Biology. She is an Elite Fear-Free Certified Professional and certified in Low-Stress Handling for Dogs and Cats, as well as certified in small animal massage therapy.

In addition to this book, Alana is the author of *Training Your Dog the Humane Way* (2011). She is the creator of the webinars; Feline Fundamentals: Humane Handling Basics and Behavior, Body Language and Communication, and Canine Fundamentals: Behavior and Body Language. Alana created standard operating procedures (SOPs) for animal shelters on Humane Feline-Handling and the Use of Dog Harnesses, and she is on the Cat-Friendly Practice Advisory Council for the American Association of Feline Practitioners.

She can be reached through her website *AlanaStevenson.com*.

1

KITTEN BASICS

KITTENHOOD

Just as humans go through developmental changes with age, so do other animals. Human babies will start to crawl, walk, stop nursing, pick up language, and learn to speak at specific times or windows in their development. All baby animals, including kittens, go through similar developmental stages. These are called 'sensitive periods.' Exact timing of these sensitive periods varies between individuals.

Missing a 'sensitive period' does not mean an animal will be poorly socialized or will not be able to learn at a later time. Later learning is still important. During the sensitive period, however, their body is naturally receptive and ready to learn from certain changes and exposure. Observing and learning from the behavior of other cats, especially the mother, is especially significant.

There is a sensitive period for socialization called the *socialization period*. This is different from the word socialization as we frequently use it. The socialization period is a stage in the kitten's development when fear is relatively low and the kitten can recover quicker or bounce back faster from potential stressors. This 'socialization period' or 'sensitive period for socialization' occurs between 3 and 9 weeks.

KITTEN STAGES

The *Neonatal period* is the first two weeks of life. One-day-old kittens can detect heat and cold and will approach warmth. They are born with a sense of smell, but their ears and eyes are closed. Kittens have a suckling or 'rooting reflex' and will initiate suckling on the mother by gravitating to particular nipples. The preferred nipples are the ones closest to the mother's face. Female cats have eight nipples, but only six produce sufficient milk.

The first milk the mother produces is the colostrum. This is rich in antibodies and helps to protect kittens against disease. These antibodies last for about six weeks. Kittens can only absorb colostrum during their first 16–24 hours of life and should be nursing within two hours of birth. It is essential that kittens receive colostrum to protect them against diseases.

DOI: 10.1201/9781003351801-1

1–3 WEEKS

Kittens' eyes begin to open between 7 and 10 days. Though kittens can hear after the first week, their ears open at two weeks. At three weeks, they begin trying to walk and their ears become upright. They can recognize their mother by sight and smell, and they paw paddle and purr while nursing. At this age, kittens cannot eliminate on their own. The mother cat stimulates her kittens to eliminate by licking their genital area.

During the first three weeks of life, the mother and kittens communicate to each other through ultrasonic calls. After kittens' eyes open, they see other animals in the nest who they consider littermates. Social contact with the mother is especially important during this time, and the mother regularly licks and nuzzles her kittens to nurture and calm them. If the mother is outside and free-roaming, she moves her kittens whenever the nest gets dirty or disturbed.

Kitten calls

By three weeks, there are two calls produced by kittens: purring and distress. Purring occurs when they are nursing. Purring by the mother and littermates calms kittens and keeps them together in the nest.

The distress call is when kittens are isolated, cold, or trapped. The isolation call is usually the loudest. Deaf kitten calls tend to be louder or lower pitched than their siblings who hear. When the kitten is lost or alone, she will call in distress and the mother will come to her aid.

Other than purring, when kittens are nursing, they should be quiet. If they cry, it means they are not getting something they need or may be sick.

The full meow is not developed until about 11 weeks.

At **4 weeks**, vision develops, self-grooming begins, and kittens can retract their claws. They begin to preamble, play with littermates, and can recognize familiar voices. Like most babies, kittens greatly prefer their mother over other adult cats.

By **5 weeks**, hearing becomes fully developed. Kittens start batting at objects and playing by themselves. They start climbing, scratching on surfaces, and begin to eliminate on their own. The mother will still stimulate elimination by licking the genital region. Predatory or hunting behavior also begins at this time.

By **6 weeks**, kittens should have voluntary control over elimination. They start to dig and cover. Their baby teeth emerge and their eyes become clearer. Vision develops. Self-play increases, including playing with their mother's and littermates' tails. For free-ranging cats who hunt for food, the mother starts to bring kittens live prey so they can practice hunting skills. Weaning may begin.

At **7–8 weeks**, kittens can finally regulate their own body temperature. Before this time, they are unable to so they need to be kept warm. Social play continues with wrestling, rolling, biting, and chasing and eye-paw coordination develops. This is the time play is directed toward human hands and feet, as well as moving objects. Kittens begin to show adultlike behavior to social situations and stressors. Most kittens will still be nursing on their mother.

Between **8 and 12 weeks**, weaning is completed, and by 12 weeks, the mother's milk quality also changes. However, many kittens still nurse or try to nurse at this age and the mother may allow it. Kittens lose their baby teeth and adult teeth emerge. They should have all their adult teeth by the age of 6 months.

Between **9 and 16 weeks**, kittens will be eating solid foods, and if they are outdoors, they will go exploring and hunting with their mother.

Food preferences

At about 5–8 weeks old, during weaning, kittens develop their mother's food preferences. If the mother prefers certain types of food, they will too. This can even occur with novel foods that wouldn't normally be part of a cat's diet. In one study, mother cats were taught to eat bananas or mashed potatoes. Their kittens primarily chose the banana or mashed potato over meat pellets (Wyrwicka, 1978; Wyrwicka & Long, 1980).

SOCIALIZATION PERIOD

The *socialization period* is between 3 and 9 weeks. This is the 'sensitive period' for socializing kittens to other animals, including humans. Since kittens are receptive to people and other animals at this time, gentle and frequent handling is important. Socialization means positive exposure to other cats, people, and animals at the kitten's individual comfort level. Socialization is not just 'exposure' itself.

Kittens who are gently and frequently handled by multiple people will be friendlier later on than kittens handled by only one or a few individuals. If a kitten isn't exposed to gentle handling and touch during the socialization period, she will be more fearful, defensive, or avoidant of people and will likely not initiate interactions or approach them.

If a kitten has not been exposed to other animals during her socialization period, she will be more fearful or defensive upon meeting them. If a kitten hasn't been positively exposed to other cats by the time she's 10 months old, it will take a bit longer for her to accept them.

Observational learning also affects kitten behavior (Figure 1.1). If the mother is friendly and approaches people or other animals, her kittens will too. The presence of the mother and siblings boosts a kitten's confidence. They approach new and novel objects more quickly as well.

NATURE VS. NURTURE

Although the amount and quality of handling are important for how social a cat is, genetics has an effect too. If a cat is genetically predisposed to being timid or less friendly, they may not respond to handling as well or to the same extent as non-timid cats.

Friendliness to humans is partly influenced by paternal genetics. It's been shown that the temperament of kittens may be largely determined by the father. Interest in approaching unfamiliar objects can also be related to paternity. Cats fathered by 'friendly' males are more likely to explore unfamiliar people and inanimate objects than cats fathered by timid or less outgoing ones.

Figure 1.1 Kittens learn from observing their mothers. © Creative Commons/Dreamstime.jpg.

Both socialized cats and cats with a friendly father are friendlier to unfamiliar people and tend to be less distressed when people touch them. Littermates of friendly fathers have similar behaviors which indicate an inherited component to behavior, as well.

MALE VS. FEMALE KITTENS

Before 8 weeks, male and female kittens are similar in weight and behavior. After 8 weeks, males tend to be larger. At 3–4 months of age, differences in play styles emerge. Males tend to wrestle more and are more aggressive when playing than females. Females prefer to stalk or hunt inanimate objects, strings, and toys. Females who play with males become rougher when they play than females who play with females.

SEXUAL MATURITY

Male cats reach puberty or sexual maturity at about 9–12 months. Sexual maturity in females is between 6 and 9 months of age. However, a first heat cycle in female kittens and spermatogenesis for male kittens can occur as early as 5 months, so early spaying and neutering is recommended.

SOCIAL MATURITY

Social maturity for cats is 3–5 years for males and 2–4 years for females. Males reach adult weight at about 3 years old. Females take about 2 years. We often think a 2-year-old cat is an adult because they are larger than kittens, but most have only begun social maturity. Like humans, social maturity varies between individuals.

The longevity of a cat is 16–25 years.

MATERNAL BEHAVIOR

Intact female cats have two to three heat or estrus cycles per year. Estrus lasts 9–10 days. Gestation, or length of pregnancy, is about 63 days. The mother gives birth to approximately 3–5 kittens. A first-time mother or younger mother will tend to have smaller litters. Unspayed females have approximately two litters per year.

After the birth of each kitten, the mother licks away the amniotic sac and chews off the umbilical cord. When the placenta is expelled, she may eat it as nourishment so she can stay with her kittens for a few days to nurse them.

The mother cat will frequently move her kittens when the nest becomes dirty or disturbed.

In colonies of intact cats, kittens may suckle from females other than their own mothers. This often occurs among related female cats. When kittens stay with their mother, she may show maternal behavior to them even when they're adults.

If there are enough resources for them, female kittens will stay with their mothers. Intact young males will stay with their mother and female siblings for the first year or two, and then will start leaving the group to roam or wander.

NUTRITION AND STRESS ON MATERNAL BEHAVIOR AND KITTEN DEVELOPMENT

Poor nutrition for the mother during her pregnancy and lactation can affect the behavior of her kittens. Kittens born to malnourished mothers can have a decreased ability to learn, increased reactivity to anything new, and awkward responses to other cats. Males are more aggressive during play and females climb less, but run more. In addition, undernourished mothers show less nurturing behavior and are more irritable toward kittens.

Maternal stress also affects kitten behavior and development. Newborn kittens with stressed or anxious mothers have lower birth weights and gain weight slower than kittens with mothers who are relaxed and content.

WEANING

Weaning is the transition from milk to solid food and is done gradually. The mother does not stop nursing kittens overnight, nor does she exhibit any aggression toward her kittens. During weaning, she provides less opportunities for them to nurse by standing up and walking away or adjusting her position to prevent access to her nipples. It is not uncommon for 8–10-week-old kittens to nurse. They often continue to nurse for up

Figure 1.2 It is not uncommon for mother cats to let their kittens nurse beyond 12 weeks. © Victor Guevarra.jpg.

to 12 weeks. Some mothers will still allow or tolerate their kittens nursing or suckling on them after this age (Figure 1.2).

Unfortunately, many kittens are weaned too early. Six or 7 weeks is too young for a kitten to be removed from its mother or separated from other kittens. Kittens who are weaned or separated from their mother and other kittens too soon show earlier signs of predatory behavior and have an increased tendency for play aggression.

Late weaning decreases or delays predatory behavior. If a kitten is weaned late and not taught to hunt by his mother, he will have poorer hunting skills or show less interest. Kittens also develop their mother's food preferences. So, if a mother and her kittens are regularly presented with food and are well-fed, it can inhibit the kittens' desire to hunt.

It's fairly easy to tell when kittens and young cats have been weaned too early. They will suckle objects and body parts such as blankets, pillows, bedding, clothing, ear lobes, and fingers, and tend to continue to do so, even as adults.

BOTTLE-FEEDING KITTENS AND KITTEN CARE

If you have a newborn kitten who was orphaned or kittens you want to bottle-feed because they were removed from their mother, there are important things to know.

Kittens cannot regulate their body temperature until about 7 weeks old. Providing heat and warmth for kittens is vital to prevent hypothermia. A heating pad should be placed under towels or soft fleece for them to access. The heat source shouldn't be too hot (don't place kittens directly on a heating pad) and they should be able to move away from it when they want to.

When you first get an orphaned kitten, she needs to be warmed up before you feed her. This is important for all baby animals. Once she is warm, you can begin bottle-feeding. It's always good to provide a little sugar water, electrolytes, or Pedialyte to prevent hypoglycemia before feeding. You can also rub corn syrup or Pedialyte on their gums. Newborn kittens should be fed every 2–2.5 hours so be prepared to be busy and to get up multiple times at night.

Hygiene is critical. If a kitten hasn't received colostrum from her mother, she has no immunity. Colostrum is obtained within the first 16–24 hours of birth. Full immunity from colostrum doesn't occur until about six weeks. Since kittens don't have strengthened immune systems, they are easily susceptible to getting sick or becoming ill. Wash your hands before handling kittens, use new clean cloth or cotton balls when stimulating them, and thoroughly clean all syringes, nipples, and bottles before reusing them. After washing syringes and nipples, place them in boiling water for a few minutes to sterilize them.

Kittens cannot eliminate on their own until 4–5 weeks of age. You will have to stimulate kittens to empty their bladder and bowels. This should be done before and after each feeding (Figure 1.3). To help a kitten eliminate, very gently tap or wipe a warm moist cotton ball on the genital region. Urine should be colorless or light yellow and feces will be pasty and yellowish or tan.

Newborn kittens are quiet most of the time and should sleep, only waking to eat. Kittens will cry if they are hungry or cold. When young kittens cry out, this is a distress call and needs to be taken seriously and immediately addressed.

Ideally, kittens should be fed a commercial kitten milk replacer. KMR has been causing deaths in infant animals so I do not want to recommend any brands as this time. The

Figure 1.3 Gently stimulate kittens before and after each feeding. © Vladans.jpg.

milk should be warmed to approximately 35°C (95°F). Do not give kittens milk that is too hot or too cold. Test the temperature of the milk before giving it to the kitten by letting it drop on the inside of your wrist. If it is the right temperature, you shouldn't feel it.

If you don't have immediate access to pre-made kitten formula, human baby formula can be used in an emergency. You would double the recommended concentration of human baby formula to feed kittens since kittens require twice as much fat and protein as human infants do.

Cow's milk is suitable for calves, but not for kittens. Lactose, calcium, and phosphorus levels are too high in cow's milk and the protein and fat contents are much too low. There's less lactose in human infant formula than cow's milk.

You can find recipes online for homemade kitten formulas through veterinary and cat rescue websites. Hoskins kitten formula is a standard go to. Keep in mind homemade kitten formula is for temporary use, so it is best to get your hands on pre-made kitten formula as soon as you can.

When choosing nipples for bottles, smaller nipples are better. Many nipples on store-bought bottles for puppies and kittens are too large for tiny kittens. If the nipple is too big, it is difficult for the kitten to suckle. Often, small 1–3 cc needleless syringes with soft nipple attachments such as the Miracle Nipple, made for squirrels and other small animals, or Bubble milk silicone nipples are easier for nursing small kittens. The flow of milk should be 1–2 drops per second.

FEEDING POSITION

Kittens should nurse on their stomachs with their head slightly elevated (Figure 1.4). This is the way a kitten would naturally nurse on their mother. Do *not* turn kittens upside down or onto their back to bottle-feed them. Although this may look cute and is what we do with human babies, it can cause kittens to aspirate. They are more likely

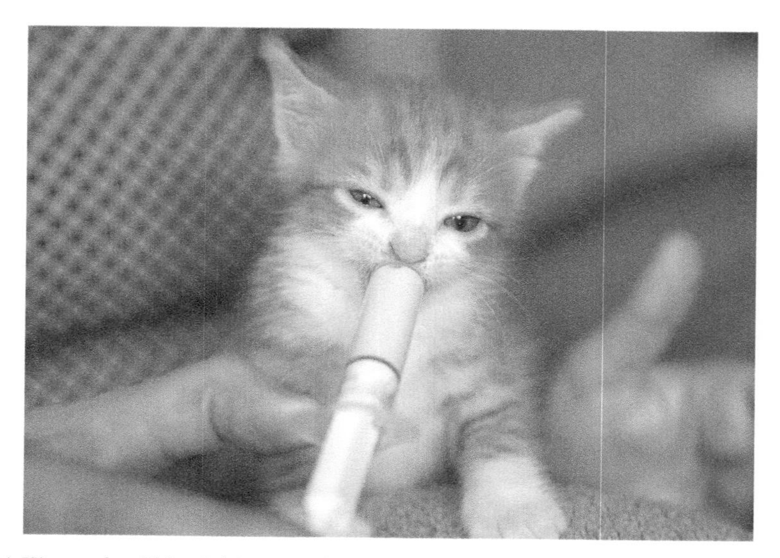

Figure 1.4 Kittens should be fed in a comfortable position on their stomachs, fully supported, and with their head only slightly elevated. © Raquel Pedrosa.jpg.

to inhale the milk and develop pneumonia. Ideally, kittens should be wrapped in fleece or a small blanket while you nurse them to keep their body temperature up.

When the mother nuzzles the faces of her kittens, it calms them. You can clean, groom, or nurture kittens as their mother would do by softly wiping their head, face, and chin with a warm moist cotton ball. Dab the cotton ball in warm water and squeeze it out so it isn't wet and doesn't drip. Then, follow with a dry one.

When kittens are about 4–5 weeks, provide a shallow litter pan for them filled with wheat or corn litter. Clay and clumping litters can be used when you know the kitten will not play with or ingest litter.

Weaning can begin at 4–5 weeks and is a process that should take several weeks. Weaning should not happen overnight. Kittens should be able to nurse from the bottle until they choose to eat on their own. To begin weaning the kitten, put formula in a shallow bowl or on a spoon and encourage the kitten to lap it up. When your kitten starts to lap the formula, mix wet food with it, but so it's still a liquid. As the kitten becomes more comfortable lapping on her own, add more wet food to the formula until you can wean off the formula and she can eat the wet food. Since early-weaned and bottle-fed kittens tend to develop earlier predatory behavior, I would personally allow kittens to nurse from the bottle for as long as they want or until they choose not to.

Since kittens are so vulnerable and have little immunity, it's important to seek out your veterinarian for any questions you have about the kitten's health or proper kitten care. If a kitten is not doing well, does not seem to gain weight, cries frequently, or seems in distress, she needs to be seen by a veterinarian.

HOW TO HOLD AND HANDLE KITTENS

It's important to be supportive of kittens. The kitten's entire body should be supported when you hold, handle, or carry him (Figure 1.5). This includes his chest, torso, abdomen, and hind end. If the kitten is very small, support his limbs as well. When you place kittens down, do so gently. It's easy for us to pick kittens up by the middle and let their little legs dangle, but this is not pleasant or healthy for them. Kittens will become more wiggly next time and more apprehensive or fearful of handling.

At no time should a kitten be 'scruffed' or held by the back of the neck and dangled in mid-air (Figure 1.6). This type of handling startles and frightens kittens. They usually squirm or cry in response.

Mother cats carry newborns in their mouths because they have no choice. The mother has good intentions – to move her kittens to a safer place. Mother cats are very sensitive to how much pressure they use when they carry their kittens. They do not scruff their kittens to interact with them socially or to punish or correct them.

THE IMPORTANCE OF SPAYING AND NEUTERING

Although not common, female kittens can go into heat or estrus and boy cats can produce sperm as early as 5 months of age. For females, a heat cycle will last about 10 days. At this time, she will cry, wail, pace, and raise her hindquarters. She may also bleed. If she is outside, intact males will fight with each other and try to mount her.

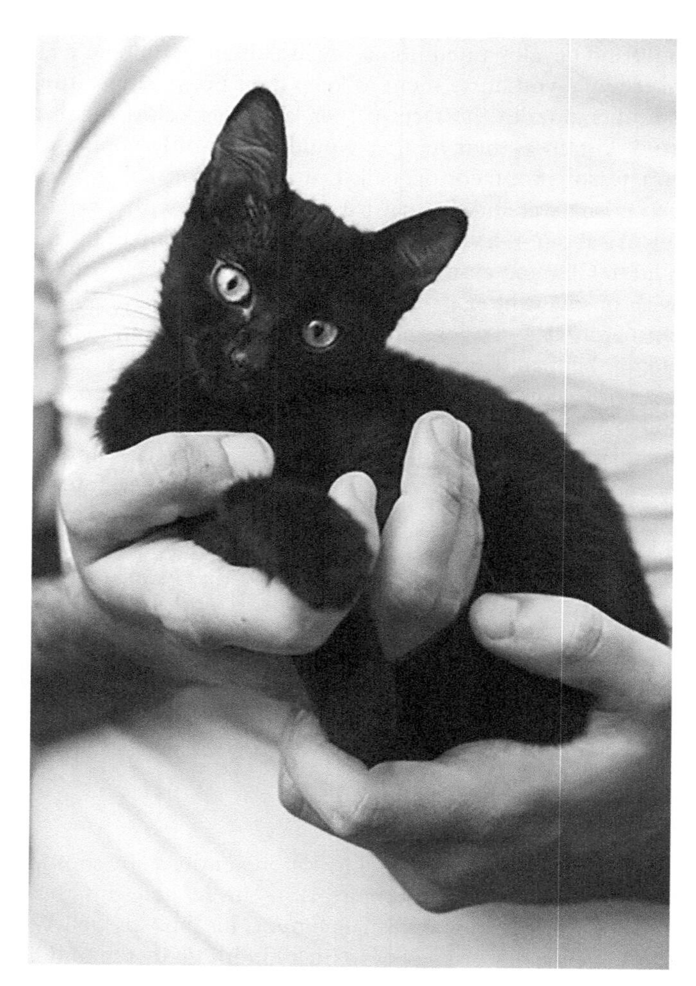

Figure 1.5 Kittens should always be held and carried with their body fully supported. @ Catalina Zaharescu.jpg.

This is one reason you should spay or neuter your cat as early as possible. Not only is the heat cycle stressful for the cat, it's also stressful for the person who has to live with her. There's no reason to let your cat have unwanted pregnancies.

In my opinion, six weeks is unnecessary or too early to spay or neuter a cat, but three to four months is not. The early spaying and neutering of kittens who are sexually immature does not cause any health problems, nor does it negatively affect their behavioral development. There is no reason for your cat to experience her first heat before she is spayed. It is unnecessary and it's uncomfortable for her.

Spaying and neutering cats eliminates their risk of mammary, ovarian, uterine, and testicular cancer, as well as pyometra (infection of the uterus).

Early neutering of male cats eliminates the likelihood that they will urine mark or spray. Young male cats may begin to spray at one year's old. Early neutering

Figure 1.6 At no point, should a kitten be carried by the 'scruff' or dangled in mid-air. @ Pranav Kumar Jain.jpg.

significantly decreases roaming and fighting, along with injuries and abscesses associated with fighting, and diseases such as Feline Immunodeficiency Virus.

A spayed or neutered cat does not contribute to pet overpopulation. Overpopulation means that there are too many cats for the number of good homes available or people who want them.

Intact male cats can make very good housecats once neutered and can be particularly affectionate.

FERAL OR STRAY – WHY IT MATTERS

There is often confusion in the use of the words feral and stray. It is important to know or be cognizant of the differences since 'feral' is a word that is frequently misused and overused. The label can have dangerous consequences for cats since most shelters will kill 'feral' cats at intake, or people assume the cats are behaviorally damaged and can never be socialized to people. There is a stigma associated with feral cats.

What does *feral* mean? Put simply, a feral cat is a cat who has never interacted with or been positively handled by people during the socialization period in kittenhood. This is different from a stray or abandoned cat. Many outdoor cats in the city or suburbs are not feral, but strays. These homeless or outdoor cats were once owned by someone, even if it was for a short time. The cat rummaging around the garbage bin in an alleyway or behind someone's yard may or may not be feral. Most likely they aren't, regardless of appearance. They may have been abandoned, lost, or on their own for a long time. A free-roaming or outdoor cat might be feral, neglected, or simply an indoor–outdoor cat allowed to wander.

Homeless and stray cats initially can be quite frightened of people even though they were handled as kittens. Cats who are bonded to only one or a few individuals, when away from their owners or special people, will be terrified of new people or when placed in new environments. They will panic and hide. When cats are caught in traps, they often panic or may respond aggressively. Some cats can remain in this emotional state for days or even weeks afterwards. Their behavior will seem no different than that of a feral cat.

There is no way to know if a cat is feral unless you are aware of the cat's history in kittenhood prior to 9 weeks of age. You cannot tell if a cat is feral by their initial behavior or by looking at them (Figure 1.7). Only spending time with the cat and allowing the cat to acclimate in a calm, safe, quiet environment will shed light on the cat's prior socialization experiences.

Some individuals prefer terms such as 'free-roaming' or 'community cats' when labeling outdoor cats. Language is important. I do believe it is damaging to cats when we instantly label them as 'feral' because they were found outdoors, are fearful, or look neglected and uncared for.

Figure 1.7 You cannot tell if a cat is feral just by looking at him or his initial behavior. A clipped ear simply means the cat has been trapped, sterilized, and released. © Julia Igosheva Kowlas.jpg.

Feral cats can acclimate to one or two individuals over time, although they are generally not as affectionate and it can take a long time to get them acclimated to people, movement, and touch.

Unfortunately, stray and feral cats don't fare very well on their own. Without assistance from people, safe shelter, and food, the average life expectancy for stray and feral cats is 3–5 years.

REFERENCES

Wyrwicka, W. (1978). Imitation of mother's inappropriate food preference in weanling kittens. *Pavlovian Journal of Biological Science*, 13, 55–72.

Wyrwicka, W. & Long, A. M. (1980). Observations on the initiation of eating of new food by weanling kittens. *Pavlovian Journal of Biological Science*, 15, 115–122.

2

BASIC FELINE CARE & ENRICHMENT

FEEDING

Cats are carnivores, so they can't move their jaws laterally the way we can. This means they can't chew or grind their food as we do. Their jaws move primarily up and down, and their premolars are serrated, not flat, to cut through flesh. This is why cats eat by gulping or taking chunks or pieces of food and swallowing it. While most mammals, including us, can taste sweet, sour, salty, and bitter, cats have very few sweet receptors on their tongue, so they don't taste sugar or sweet foods the way we do.

Cats do not like cold food. If you warm your cat's wet food when it comes out of the refrigerator because you think your cat prefers it, you're right! Preferred food temperature for cats is average body temperature, 37°C (98.6°F). This is because they evolved to eat prey animals. Mice have a body temperature of 36.5°C–38°C. A rat's temperature ranges from 35.9°C to 37.5°C, and a bird's can be 39°C–43°C.

Cats naturally can eat up to 20 small meals a day. They really are nibblers and grazers, eating small amounts frequently throughout the day and evening. Leave dry food out for cats to nibble at will. Most cats are better behaved and more relaxed when they are free-fed (Figure 2.1). See 'Food for Thought,' p. 156. If your cat enjoys wet food, provide that as well. Wet food can add moisture to your cat's diet. (Try to avoid wet food with artificial colors.) If you want to free-feed your cat wet food, you should wash the bowls and freshen the food every 3–4 hours. If you have multiple cats and feed them meals, every cat should have their own bowl or plate.

It's important to freshen the food and wash plates and bowls daily. If wet food sits out in a bowl, it becomes rancid quickly, within a few hours. Dry food, depending on temperature and humidity, will start turning stale too. When cat food is left out uncovered for too long and bowls are not washed, it attracts flies. When that happens, cats don't want to eat it.

Although it's currently the trend or popular to feed house cats little portions one to three times a day, cats who are put on meal restriction and fed rationed portions are more irritable and aggressive than cats who are free-fed. With multiple cats, this frequently becomes apparent before feeding time. Meal restriction and hunger increase the likelihood, intensity, and frequency of play aggression, petting aggression, and inter-cat aggression. It can also cause a cat to excessively vocalize, especially at night and in the early hours of the morning.

DOI: 10.1201/9781003351801-2

Figure 2.1 Cats do better behaviorally when they are free-fed. © Gaelle Marcel.

In addition, restricted meals or diets can cause cats to become obsessed with food, so when they do eat, they gorge. When cats eat too much or too quickly on an empty stomach, they frequently vomit. This can cause people to worry that their cat is sick or allergic to certain foods, when their cat is simply hungry and eating too fast. If your cat seems ravenous and is raiding the refrigerator, begging you for food while you're eating, or trying to steal food from your plate, she is hungry and needs to eat more.

Kittens and young cats should eat when they are hungry and until they are satisfied. Young animals have growth spurts and developmental stages they go through. At times, they will eat more, and at other times, they will eat less. It's important that young cats and kittens are not deprived of food or kept hungry.

Cats should not go without food for longer than 24 hours, even less for kittens. Never starve a cat so she eats a particular type of food. It is better and healthier for the cat to eat whatever she likes than to eat nothing at all.

Cats have a 'primordial pouch' which is an excess layer of fat and skin on their lower abdomen. This skin can sag or hang down looking like a sagging or fat belly. Even skinny cats have this pouch, and for some cats, it's more pronounced than others. Don't confuse the 'primordial pouch' with obesity or your cat being overweight.

If you must restrict your cat's calorie intake, change the types of food you feed her. Switch to a lower calorie food and increase your cat's exercise through interactive play sessions or by adding vertical surfaces and platforms for her to climb on.

Don't mix wet food with water in hopes it will make your cat feel full. This is fine if you want to add water to her diet, but not for weight loss. When a cat's meal is diluted with water, she compensates by eating more. Low-calorie foods specifically for weight management or weight loss have ingredients such as kaolin and cellulose which make the cat feel full, even though she's consuming less calories.

Some people may have heard that cats can't eat tuna. Fish and tuna are fine to feed your cat. However, if a cat is fed solely a raw fish diet, she can develop a thiamine deficiency. Raw fish contains the enzyme thiaminase, which destroys thiamine.

WATER BOWLS AND PLAYING WITH WATER

Cats often don't drink enough water. One of the reasons is cats prefer to drink away from where they eat, and we tend to keep food and water bowls next to each other. Cats in the wild drink water in locations separated from their food. This is why cats will prefer to drink from sinks, mugs, and drinking glasses, or the bathtub after a shower. The more water cats drink, the healthier they tend to be, the nicer their coats, the less they shed, and the less prone they are to developing urinary tract infections.

Large ceramic salad, soup, and serving bowls with wide mouths, as well as medium-sized ceramic dog food bowls, work well as water bowls. Many cats like running water, so providing cats with fountains is a good idea. Fountains should be wide-mouthed and easily cleanable.

Water bowls are best placed near sleeping or social areas for the cat such as in rooms they frequent, bedrooms, or near areas they sleep, and near cat trees and window perches. If you keep food and water in the same room, place some distance between them instead of positioning the bowls flush next to each other. Keep water bowls away from the litter pan.

Cats like to position themselves so they have a nice vantage point or landscape view of the rooms they are in. So, place a water bowl a few inches away from a wall or in a position where the cat has a nice vantage point of the entryways.

Young cats, especially, like to play with the water in their bowls. This is more likely to occur when the bowl is small and metal since there is reflection from the metal and the metal makes a 'tinging' sound. If your cat is playing with water for entertainment, depending on his age, he will likely grow out of the behavior. If your cat is tipping the water bowl over or spilling water on the floor, get a larger, heavier bowl or place a decorative tray or mat underneath the water bowl. If you allow your cat to drink from faucets, sinks, and bathtubs, she can play with the water without making a mess.

Wash and refill water bowls *at least* once a day. No one likes drinking stale water, and sitting water attracts bacteria and becomes slimy.

ACTIVITY

Cats are especially active at dawn and dusk. You may have noticed your cat wakes you up at 3:30 or 4:30 in the morning. This is normal. I call this the 'witching hour.' Cats are extremely active in the wee hours of the morning. This is the time when people usually want to sleep in. It also coincides with the time when birds begin to chirp.

Young kittens have tons of energy. If they are not sleeping, they will be running, wrestling, chewing, and very active. As they become older, they will sleep more often and for longer periods of time, especially in the day.

If you let your cat share the bedroom with you, over time, he will coincide his daily activity and sleeping patterns with yours. When he is alone, he will likely sleep. This can make it challenging when you work all day, only coming home to eat and sleep. When you are away, your cat's been sleeping, and when you come home to go to bed, he wants to be social and interact.

Just like us, cats dream. When they sleep, you may see their eyes or paws rapidly move and their tails twitch. This is the REM state or deep sleep. Cats spend about a third of their sleeping time in this state.

HOW'S THE VIEW? LOCATION

Location is very important for cats. Where they place themselves and how they use resources is significant. Cats like to have a nice vantage point or view of the rooms or areas they are in. They want to see all entries and exits. This is why cats will sit in the middle of the hallway or stairwell. They are not purposely trying to block space; they just have a great view with multiple exit plans or escape routes if, and when, they want one.

We tend to place items and furniture against walls and in corners. Cats like to be centerpieces. This is important when positioning anything designated for your cat, whether it's a scratching post, cat shelves, litter pans, or food and water bowls. If you place a scratching post in a corner, so your cat has to turn her back to the doorway or where she might be unexpectedly startled or surprised when you enter, she won't use it. She'll use the mat, rug, or arm of the sofa that provides a nice visual of the room. Likewise, if you have a fabulous cat tower, but in the wrong location or against the wrong wall, she won't use it either. Sometimes, moving cat furniture and resources 6–12 inches away from a wall or corner makes it much more appealing for the cat.

Windows tend to be in good locations. If you put your back to a window and look into a room, usually you have a very nice vantage point, so make use of window space for cats.

VERTICAL TERRITORY AND CLIMBING

Think vertically to enhance your cat's well-being. The more vertical territory you have for your cat, the happier she will be. Vertical territory refers to platforms and surfaces for cats that are higher up and off the ground (Figure 2.2). Access to vertical, as well as three-dimensional, space can be much more important for cats than horizontal.

Cats like to climb, especially when they are young and agile. Being higher up for cats provides them with a good view of the area or rooms they are in. Cats who are prohibited from climbing on furniture or jumping on counters are often more anxious and vulnerable than cats who have multiple surfaces to climb on. When cats are

Figure 2.2 Cats like to have a nice vantage point or view of the rooms they are in. They like to climb, so they need access to vertical surfaces and platforms. © John Birkholz.

relegated to the floor or punished for climbing, they become less confident and more skittish. Cats who are under-stimulated and not allowed to climb are more likely to pace and vocalize than cats who have access to vertical territory. Not having enough vertical space increases the likelihood of play aggression, such as attacking your ankles or feet when you walk.

In multi-cat households, if there are no vertical surfaces for cats to climb on, they frequently collide with each other or have to pass each other in close proximity to access resources. If there is any tension or antagonism between cats, a victimized cat will have to run away or underneath furniture to avoid conflict. A victimized cat will retreat into closets, under sofas and beds, or in basements to avoid other cats and to feel more secure. Having vertical surfaces and different levels for cats to rest on, in the right locations, increases the overall space for cats and enables them to view their environment in safety and with confidence.

Unfortunately, many platforms, condos, and cat trees marketed for cats are too narrow, steep, or flimsy for them to easily maneuver or climb on. This is one reason cats prefer kitchen islands and dining room tables. These surfaces are sturdy and large, usually centrally located, and provide a great peripheral view for the cat.

When adding cat shelves, condos, and platforms, make sure they are easily accessible to your cat. They should be large enough for her to stand and lie down on and there should be multiple ways up and down. Most cats are about 12″ tall and 18″ long, not including their tails. So, platforms should be wide and long enough to fully support them.

About 60% of a cat's body weight is supported by her front legs. The front claws can retract and hold onto objects and surfaces. It's easier for cats to access and climb on higher surfaces, but much trickier for them to get down. Sadly, this is why cats get stuck in trees. Although cats can jump or leap down from higher surfaces, they prefer to walk or hop down, so provide them with intermediate platforms so they can easily get down from higher shelving and furniture.

PLANTS

It is perfectly fine and healthy for cats to eat grass. Wild and feral cats frequently eat it. Grass and edible or cat 'safe' plants can be placed in different locations to provide mental and physical stimulation for them.

If your cat is craving foliage and greens, allow her to indulge. Just be sure the grass she is eating has not been chemically treated. Cats tend to prefer crab grass, meadow grasses, and winter rye grass over store-bought wheat grass or 'pet-grass.'

Vomiting after ingestion of grass and edible plants is normal. Many plants are anti-parasitic which is one reason cats may ingest them. When cats throw up after eating grass, they are generally unfazed. It takes longer for us to recover. Plants also provide roughage and fiber, which is cellulose. Even though cats are carnivores, they benefit from this in their diet.

Many cats love catnip. Catnip is native to Britain and grows wild in Europe and North America. It's a perennial and is easy to grow if you have a yard or patch of grass. Young kittens tend not to respond to catnip until they are 9–12 weeks old or once they are weaned. Whether or a not a cat responds to catnip is genetic. Some cats don't respond to it at all.

Catnip contains compounds that are mood elevators which is why cats have a good time when exposed to it. However, the affect doesn't last long. After exposure, the effect itself usually lasts no longer than 15 minutes. Cats will not react to catnip again for at least an hour. Although cats can get really revved or ramped up when they smell the plant, if they eat it, it can have a calming effect.

Cats who aren't responsive to catnip may show interest in silver vine or valerian.

If you have plants and don't want your cats to chew on them, it's best to keep them in a separate room or place them in locations your cats can't access. Scolding and punishing your cat for chewing on plants is futile and can be especially challenging for indoor and young cats, or cats who have little enrichment.

It's a great idea to grow or purchase edible 'cat-safe' plants, especially for indoor cats. There are many plants that are safe for cats, such as spider plants. Your cat may also like greens such as arugula, pea sprouts, and different types of lettuces.

SCRATCHING

Scratching is a social behavior in cats. They scratch not only to stretch and sheath their front claws, but in anticipation of something they look forward to or when they get a boost of confidence. Since it's a social behavior, cats scratch more often in the presence of others. If you feel like your cat is always scratching in front of you, you're right. Cats will scratch upon waking up from naps, when you enter a room or come home, at feeding times, and before and during play.

Cats tend to scratch rugs and furniture near the entries to frequently trafficked rooms. So, it's important to place large vertical and horizontal scratchers in places that people frequent and that are socially relevant to the cat. Adjacent to doorways and near entryways are good locations.

Cats prefer large textured, solid surfaces to scratch on. The longer an object is scratched and the more tattered it is, the more attractive it is for the cat. All scratching pads and posts should be sturdy enough that they can take the cat's weight and don't rock, tip, or fall over easily when she uses them. They should be tall or long enough to allow the cat to stretch out and located somewhere prominent (Figure 2.3).

Figure 2.3 Scratching posts should be large, stable, and placed somewhere prominent.

Double-wide cardboard floor scratchers are attractive to many cats, relatively inexpensive, and can be found in most pet supply stores.

DECLAWING

A declaw is called an onychectomy and it's routinely performed in the United States. Cats have four toes each on the back feet and five toes including the dewclaw on their front. Usually, the front paws are declawed, but it's not uncommon for all four paws to be declawed.

Declawing is the amputation of each toe at the first joint – equivalent to cutting off the tip of the finger at the first knuckle. The entire first bone (third bone at the end of the finger) is removed. If a little too much of the toe is removed and the second toe bone or tissue is damaged, it will cause permanent damage and make it painful for the cat to walk.

Sixty percent of a cat's weight is carried by his front legs. Declawed cats are forced to walk on soft cartilage. Since declawed cats are forced to walk differently, muscles that would not be normally used are used and strained. This can cause pain to other muscles and joints, as well as back pain.

Since the cat's front claws retract and are used for gripping and climbing surfaces, declawed cats can't climb. Getting down from vertical surfaces is challenging too.

Declawed cats are more likely to urinate on soft surfaces such as carpets, clothing, and bedding due to the litter and the litter box being associated with pain after surgery.

I once had a client who declawed her two older cats upon the recommendation of her veterinarian. She was told it was an easy procedure. She was distraught because two weeks after the surgery, her cats were still not using the litter boxes. Prior to the surgery, they had no issues. When I was present, I could see one of the cats, clearly in pain, lifting her front paws to avoid putting pressure on them. Unfortunately, once the surgery is performed, it's not reversible and can't be undone.

SELF-GROOMING

Cats lick, clean, and groom themselves regularly using their tongue and teeth. You'll see cats groom and clean themselves after finishing a good meal and before or after napping. Grooming can be a displacement behavior, too. For instance, if your cat falls after trying to jump on a counter or ledge, she may look around the room and groom herself.

Cats cool themselves by licking their fur through evaporative cooling. Grooming is also important for hygiene, flea prevention, and keeping the fur aligned and free of mats. Cats who are prevented from grooming or groom less are much more susceptible to flea infestations.

Cats use their front paws and forelegs to groom areas of the body they can't reach with their tongue. They will lick the front paw and then wipe their head, face, and ears. Areas that are hard to reach with their tongue and front paws are scratched with their back legs. Cats use their teeth for pulling out burrs and tangles and when cleaning their paws, nails, and in between the toes.

Happier and healthier cats groom themselves more often than cats who are fearful, stressed, or not feeling well. Cats who are forced to live in unsanitary or overcrowded conditions will stop grooming themselves. Elderly cats groom less because of arthritis and, as they age, salivary production decreases. Self-grooming and hygiene are also learned behaviors. Kittens who were not well-cared for or had mothers with poor grooming habits tend to groom themselves less.

HAIRBALLS

For many cats, vomiting hairballs is normal. The cat's tongue has spiny papillae made of keratin which hair sticks to. The hair is eventually swallowed. It is then vomited or defecated out. Cats who have long hair and groom themselves regularly are more likely to have hairballs than cats with short hair. Excess shedding occurs if the cat is ill or in poor health, in dry heat, when cats are dehydrated, and in the spring when weather gets warmer. The frequency of hairballs may be worse at these times.

If your cat has long hair or is prone to hairballs, it is helpful to regularly brush and comb her to remove excess hair and keep her free of matts and tangles. For certain long-haired cats such as Persians, daily brushing and combing is necessary. It is difficult for them to groom themselves and detangle matts due to their abnormally dense fur.

When brushing your cat, use a soft bristled brush at first or a small flea comb. Small flea combs can be great for removing excess hair. It's important to focus on the chest, belly, and abdominal regions since these areas can get easily matted. Once your cat becomes comfortable and enjoys being brushed, you can progress to larger brushes or deshedding tools. Most cats dislike brushes with sharp, spiked metal bristles.

After brushing your cat, take a damp paper towel or wash cloth and stroke her along her back, body, and face. The damp paper towel will remove excess hair that combing and brushing left behind.

Cats who don't drink enough water or have poor diets can have rougher coats and shed more. Adding water bowls to your cat's sleeping and social areas, away from her litter box and food, can encourage her to drink more.

Giving your cat store-bought hairball gels or a little fish or coconut oil on a regular basis may help as well.

If vomiting hairballs becomes too frequent or if your cat seems ill, it's important to get her checked by a veterinarian.

ELIMINATION

Cats prefer to urinate and defecate in different locations or litter boxes and avoid eliminating where they eat, drink, or in their social areas.

Since cats are obligate carnivores, they excrete ammonia directly instead of it being diluted and processed into urea by the kidneys. This is why a cat's urine has a strong ammonia smell to it.

Cats generally squat when they urinate, but some cats, if there is a vertical surface, will urinate while standing. They may raise up on their hindlegs and urinate on a wall or post. This is not the same behavior as spraying, as usually the cat will then scratch the surrounding area to 'cover' the urine.

If your cat usually squats to urinate, but all of a sudden urinates or tries to urinate while standing, see the veterinarian. Neutered male cats, especially, can exhibit this behavior when they have a urinary tract infection, bladder stones, or crystals.

When cats urinate and defecate, they find a spot to dig a hole. They then eliminate and cover their waste. When cats cover their waste, they usually make a full circle, covering it from multiple directions by scratching the ground and soil surrounding it. This generally means that cats utilize a space of approximately 2 × 3 ft when eliminating outdoors.

All cats are individuals, but a healthy, well-fed house cat should urinate 3–6 times per day and defecate 1–2 × per day.

HUNTING

Cats primarily hunt small mammals such as rodents. Their hunting strategy is to hide, wait, and ambush. After waiting for a period of time, when the opportunity presents itself, the cat will pounce three to five times and either catch the prey animal or the prey gets away. The chase is usually fleeting and short-lived.

Cats coincide their activity with the activity of their prey. This tends to be at night, dusk, and dawn. In hotter weather, cats will hunt in the evening, and in colder weather, they will hunt during the day.

Cats typically eat up to 20 small meals in a 24-hour period. This means they have to hunt for many small critters to sustain themselves. Because hunting doesn't always bring the chance of success, cats are opportunists. In other words, they will hunt whenever there is an opportunity, regardless of hunger. This is why well-fed cats will still hunt.

A cat's canines can dislocate the vertebrae and sever the spine of prey. When cats are skilled predators, they direct a bite at the back of the prey's neck. This is often called the 'nape' or 'kill bite.' Cats who hunt for survival and mother cats with kittens to feed regularly use the nape bite to kill prey quickly. The nape or kill bite is practiced regularly in kittenhood.

Not all cats are great hunters. Female cats, especially with kittens, tend to be much more efficient than males. Cats who were raised and taught to hunt by their mother will be better hunters as well. Early-weaned kittens show predatory behavior much sooner than normally weaned kittens.

Since kittens prefer prey and eat food their mother prefers, if the mother is presented food on a plate, her kittens will expect food on a plate, too. Kittens who are weaned late tend to have poorer hunting skills than kittens who are weaned early. This doesn't mean they won't hunt, but it does mean they are not very good at it.

Housecats who are allowed outdoors, have a safe place to sleep, and free-access to food when they want it tend to hunt for recreation. They are more inclined to 'play' with their prey. Instead of killing the prey quickly with the nape bite, they bring it back to the house half alive and eat their cat food, or it slips away, injured. Even if indoor–outdoor housecats succeed in killing prey, they often don't eat it.

When cats are hungry, especially in a city or areas where there is limited wildlife, they scavenge for food. There is a misconception that street cats or strays will feed themselves on rats. Cats are carnivores, but they are also scavengers. Because living on the streets is stressful for most cats (anxious and fearful cats are often not confident enough to hunt) and there is not much wildlife in cities, most street cats and strays scavenge around garbage bins, dumpsters, and restaurants or solicit food from people. Homeless cats and street cats, including cats in 'feral colonies,' tend not to fare very well without food and help from good Samaritans.

Most of the information on prey animals preferred by cats comes from studying the intestinal and stomach remains and feces of free-roaming outdoor cats. With exception of cats living on islands, cats greatly prefer rodents and small mammals over birds. Preferred mammals are voles, field voles, baby rabbits, baby squirrels, and in North America, chipmunks. Cats prefer voles over mice. Shrews are rarely found. Cats will eat insects such as grasshoppers and moths, as well as small amphibians and reptiles, such as snakes. Unlike the popular stereotype, cats tend not to eat rats.

On islands, birds are a common prey animal for cats. Cats target and prefer ground feeding birds, flightless birds, fledglings, and seabirds who nest on the ground.

INDOORS VS. OUTDOORS

Many owners keep cats entertained by giving them unsupervised outdoor time. But letting cats roam puts them at risk of getting into fights which can lead to abscesses or injuries, getting hit by a car, being stolen, poisoned, accidentally shut in a garage or shed, or picked up by someone and brought to a shelter.

In more rural areas, outdoor cats can get caught in body-gripping traps. These are still legal in many states. Cats get stuck in trees too. Although cats are good climbers, since their front claws grip onto objects and their forelegs take most of their weight, it can be challenging and sometimes impossible for them to get down.

If a cat who regularly goes outdoors and is familiar with the area disappears, she didn't get lost. Something happened to her to prevent her from returning home. Free-roaming cats and cats allowed outside unsupervised have a much higher mortality than indoor cats.

Without a doubt, cats who have access to the outdoors as a choice can enjoy it, especially if they live in suburbs, rural areas, and where there are nice lawns and gardens. Being outdoors for them is a recreational activity. It's very different for cats who have to survive outside on their own without safe access to food or shelter.

Happy well-fed housecats allowed to go outdoors hunt, but they are rarely efficient at it. Since they don't need to hunt for survival, most indoor–outdoor cats play with their prey or it slips away injured. Cats who have to survive by hunting kill quickly with a nape or kill bite. City or street cats, with limited access to safety or wildlife, mostly scavenge.

Often rescued, timid, and older cats prefer to stay indoors. They are warm, comfortable, protected, and away from things that might startle them. Most former stray and homeless cats have had traumatic experiences outdoors, so prefer to remain inside.

Indoor cats are unlikely to get parasites. They are protected from serious injury and from getting hurt. Indoor cats spend more time interacting with people than cats who regularly go outdoors because they actively seek physical and mental stimulation.

On the flip side, indoor cats can get bored easily, especially when they are young and energetic or live in small homes or apartments. If people go to work or school all day and the cat is home alone, he'll get restless. Indoor cats have a higher incidence of behavioral problems than indoor–outdoor cats such as knocking things off of counters, attention-seeking, excessive vocalization, including night wailing, and play aggression.

If you own your own property and want your cat to have access to the outdoors, you can build an outdoor cat enclosure. These are popularly referred to as 'catios.' This way your cat can have outdoor time without any risk to themselves or wildlife.

If you have a yard or quiet area, you can also teach your cat to walk on a harness. Once your cat realizes it is for going outside, providing he is not fearful, he'll likely take to it quickly. However, once he goes outside and enjoys it, he'll want more outside time. Some cats, once they associate the harness with going outside, become obsessive about wanting to go out. In addition, walking a cat is not the same as walking a dog. The cat decides where he wants to go, what he wants to do, and where he wants to lie down. Trying to coax a cat to walk for exercise is fruitless. So, if you do decide to take your cat out for a walk on a harness, be prepared to meander and stay in one spot for long periods of time.

I personally prefer keeping cats indoors or, if allowed outside, supervising them. The risks of being unprotected outdoors outweigh any benefits. My general motto for animals under my care is 'Safety first, fun second.'

BRINGING AN OUTSIDE CAT INDOORS

Many of us have experienced an unknown cat visit our property or have seen a cat outdoors who looks lost or neglected, either due to behavior, physical appearance, or being out in inclement weather (Figure 2.4). Sometimes, upon invitation, these cats waltz right into our homes. Other times, they may be too skittish to be touched or for us to get near them.

If you need to catch a cat to bring him indoors or get him seen by a veterinarian, it's best to set up a feeding pattern that is predictable. Leaving food out regularly and at a certain time in the same location will ensure he comes to eat. If the location is unsafe or where the cat may be visible to others, place the food in the bushes or in a more secure, sheltered spot. Once the cat is eating the food regularly, it's time to set up a trap.

Figure 2.4 If a cat is outside in inclement weather and hungry, the cat is homeless or needs help. © James Cheng.

Figure 2.5 I prefer box traps such as Tru-Catch (pictured) over traps such as Hav-A-Hart by Woodstream, a trap manufacturer for the fur industry.

The best traps are box traps that have two openings and are used specifically for rescue or wildlife. I prefer Tru-Catch traps (Figure 2.5). They are easy to set up and work well. I dislike the popular Hav-A-Hart traps because the company that manufactures them, Woodstream Corporation, is the primary supplier of lethal traps for the fur industry. There's no reason to support a company that profits from harming animals when there are better alternatives available. Hav-A-Hart traps are loud, bulkier, heavier, and often collapse on the sides, which are scarier to the cat.

Set the trap in a sheltered location where you placed the food. Line it with a cut yoga mat or tape paper from a paper bag to the bottom of it. If it's warmer weather, you can cover the bottom of the trap with leaves or grass. This hides the metal and makes it more inviting for the cat to enter. Use very smelly food like canned tuna or mackerel. Drizzle the liquid from the can at the entry of the trap and leave a little trail of food to the back. At the back of the trap, place the food either on the mat, paper, or a small paper plate behind or balancing on the switch plate.

It's imperative to monitor the trap. Most cats will be active at dawn and dusk, so if a cat is hesitant, he will likely use it in the middle of the night. If the trap is next to a window or within hearing distance, add some bells to a carabiner and attach them to the trap. When the trap closes, you'll hear the bells jingle.

If, after a few attempts, the cat doesn't enter the trap, use a carabiner and some ties to keep the trap door open. This way the cat can go into the trap, eat, and leave. Once you know the cat will enter the trap, you can set it.

When cats are trapped, they don't know your intention. Expect a trapped cat to be fearful, hiss, and panic. Some cats will slam their bodies against the trap trying to escape. Other cats will claw frantically at the trap doors. The cat may hiss, spit, or lunge at you when you approach the trap or pick up the handle. If you are concerned or frightened, wear gloves. Once cats figure out you aren't going to harm them, they usually calm down and their demeanors change.

Once the cat is trapped, be calm and bring him into a quiet area or sanctuary room. Have a litter pan, food, water, and a warm bed all set up beforehand. The room should be inviting and cozy. Keep food and water away from the litter box. Release him into the room and let him de-stress. If you need to transfer the cat for transport from a trap to a carrier, see p. 71.

After an hour or two, enter the room and sit down so the cat can see you (less time may be needed if the cat isn't fearful). Read or spend time on the computer so the cat becomes familiar with you. When you look at him, slowly blink and softly acknowledge him, but don't try to coax or engage him. Pair your presence with highly palatable food. When cats are scared, they will likely eat the food after you leave the room. Once the cat feels more comfortable, gradually acclimate him to your home or other family members (see 'Socializing Fearful and Frightened Cats and Kittens,' p. 94).

LOST CAT BEHAVIOR AND TIPS TO FIND YOUR CAT

Cats behave differently when displaced or in unknown territories depending on their personalities and the circumstances in which they get lost. When a cat disappears, people can make the wrong assumptions or use ineffective approaches because they don't understand how their cat may behave. By being aware of your cat's personality and the circumstances in which she escaped or disappeared, you can focus on the right approaches so you can, hopefully, find your cat and get her back.

If a cat has been strictly indoors, she will likely hide when in a new environment. This is especially true if she is shy or nervous, frightened of unexpected noises, or leery of strangers. So, if you are on a road trip or staying at a hotel and your cat gets lost, she will be hiding in the immediate vicinity. When displaced, especially in loud or noisy environments, cats can be so fearful that they hide for days and sometimes weeks. It is unlikely they will come to you if you call them, at least not right away. They might meow pitifully, but if you're in a loud environment or there are cars in the background, you may not be able to hear them. If your cat does respond to you when you call her, she will probably do so when she has more confidence or is hungry and thirsty.

When cats are in unknown areas or displaced, they tend to come out at night when it gets dark or in the wee hours of the morning. Cats who are social and not particularly fearful of strangers will hide initially, but then wander. They will be close to where they escaped or got lost from but will search for a safe place to eat and sleep and likely stay near someone who tries to interact with them or sets out food for them. If they are not frightened away from their initial hiding place, they will probably return to it periodically.

When a cat wanders, if there are feeding stations (outdoor locations where people put out food for cats) or multiple strays in the area, she will stay close to that area or will find someone who puts out food in the backyard or on a porch for her.

If an indoor–outdoor cat familiar with the area disappears or doesn't return home, she didn't get lost. Something happened to her. If she is social, she may be staying with someone else, but she may also have gotten stuck in a tree, caught in a trap, hit by a car, locked in someone's shed or garage, or driven or chased into another area.

TIPS FOR GETTING YOUR CAT BACK

Microchip your cats or any animals you have. Keep the company's contact information on file and update the information whenever you move or change locations.

Before you travel or take any road trips, get a tracking collar for your cat. Keep it on her for the road trip and until you have gotten to your destination. When you know she is safe and familiar with the new location, you can remove the collar. This way, you have some reassurance that if she gets loose or slips away, you can find her.

Don't make assumptions that hinder your chance of finding your cat. Although coyotes are often blamed for cat disappearances, the chance of your cat being killed by a coyote or another wild animal is highly unlikely. Being caught in a trap, hit by a car, stuck in a tree, or accidentally locked in a shed or garage are far more likely causes.

If you are traveling and in a new or unknown environment, stay in the area your cat got loose and start looking, and then branch out from there.

Print out individual postcard-sized 4″ × 6″ fliers with a photo of your cat and brief description of what happened. If you were on a road trip or in a busy area, contact local hotel managers and staff as well as store clerks and gas station attendants to ask them to be on the lookout.

Go within a half mile radius, door to door, and put leaflets on people's porches, front doors, and welcome mats. If you live in an apartment building, do the same. People are much more responsive when you add a personal touch. Don't be afraid to knock on the door or ring the doorbell. When you go door to door and it's personal, people become more concerned. You'll be surprised by how many people will post your cat's picture or leaflet on their refrigerator or bulletin board, especially if you bring it to them personally.

Don't post fliers on telephone poles at intersections in hopes that people in a passing car will pay attention. Even if they did, they are unlikely to stop and get the phone number. Often, writing can't be seen on fliers posted on the side of the road.

Contact veterinary hospitals and visit animal shelters personally. Ask to meet or speak with the shelter managers or kennel staff who directly care for the cats. Some shelters have cat managers and designate certain staff to certain rooms. Instead of randomly dropping off a flier or post to the front desk, bring it to those individuals who monitor the cats that come in and are more likely to know which cats may be 'in the back.' Periodically, check in with them to keep your cat on their radar.

Find out who feeds outdoor cats in the area and the people who leave food out for them. A cat colony is where a bunch of cats will congregate outdoors, usually shelter is provided, and a kind individual regularly feeds them. If your cat is not in the immediate vicinity to where she got lost, she will look for a safe location or the nearest food source. Lost and displaced cats frequently end up in these colonies.

Don't forget to look up in trees! Cats can easily climb them, but have great difficulty getting down. More cats get stuck in trees than people realize, and contrary to popular belief, they cannot get down by themselves without serious injury. If you do happen to find your cat in a tree, contact an arborist or a tree climber.

WHERE TO LEAVE FOOD AND SET TRAPS

Establish feeding stations next to the location your cat escaped from or got lost. Place food bowls in areas that are hidden or sheltered for the cat. Continue leaving food at these locations and check on the bowls regularly. If possible, purchase an outdoor camera so you can monitor the locations where you leave food. Once you see food being eaten fairly regularly and you think it may be your cat, set up a trap. Monitor and check traps, multiple times, especially at night and early in the morning.

When a trap is set with food, other animals can and will enter the trap, so it's important that you religiously monitor and frequently check on these traps. You don't want to unnecessarily frighten other animals and you don't want your cat to see other animals caught in the trap. The better you are at monitoring the traps and releasing animals who are unfortunately caught in them, the more likely you are to catch your cat.

If after a week or two, you have no success, whether from visual sightings or monitoring the feeding areas or traps, branch out another half mile to a mile radius. Pass out the notecards or fliers and go door to door. If you have friends or are able to cover more area sooner, do so and don't give up. Your cat likely wants to be back home with you.

3

BODY LANGUAGE & COMMUNICATION

EYESIGHT

Cats are capable of color discrimination and can see differences in the textures, shapes, and sizes of objects. They are very sensitive to movement, especially rapid motion, and can see well in dim light. Cats have a tapetum lucidum, an iridescent reflective layer of tissue behind their retina, that improves night vision. This makes their eyes shine or glow when illuminated. It is common in nocturnal and crepuscular creatures. Cats can see with a sixth of the amount of light that we can.

Cats have large, forward facing eyes, which gives them good stereoscopic vision or depth perception. However, breeds such as the Siamese have poorer depth perception due to poor breeding practices.

Research has shown that caged cats tend to be myopic or 'short-sighted' and free-ranging cats tend to be hypermetropic or 'far-sighted.' Since animals have to physiologically adapt to the environments they are in, this makes sense, but to what extent and how this affects indoor–outdoor cats or housecats is unclear.

HEARING

The cat's sense of hearing is far superior to ours. Cats are particularly sensitive to ultrasonic noise (sounds above our hearing range). Most rodent communication is inaudible to humans, but not to cats since rodents communicate with each other ultrasonically. Like many animals, cats have large movable ears called pinnae that swivel 180° to pinpoint the source of sound. Just as we do, cats lose their hearing as they age. Some cats, especially white-haired blue-eyed cats, are born deaf, which is genetic.

Sound is measured by both volume and frequency. Volume is measured in decibels (dB) and is how loud the sound is. Normal human conversation is generally within 55–65 dB. Those with sensitive hearing will find this too loud. A cat's loud meow or yowl is roughly 45 dB and a loud purr is about 25 dB.

Frequency is measured in Hertz (Hz) and is the pitch of sound or how high or low it is. Sounds above 20 kHz are considered ultrasonic and pitches below 20 Hz are considered infrasound. These measurements are according to our hearing – in other words, what is above or below our hearing range. Low-frequency sounds travel further

DOI: 10.1201/9781003351801-3

than high-frequency sounds. Most animals we consider silent communicate with each other regularly. We just can't hear them.

Although the average hearing range for animals, including humans, varies according to the source of sound, a cat's hearing range is approximately 45 Hz to 79 kHz. Mice generally hear sounds between 1 and 91 kHz. The dog's hearing range varies by breed but is approximately 67 Hz to 45 kHz. Although some people, usually young children and those with sensitive hearing, can hear sounds up to 20 kHz, the average adult can't pick up sounds higher than 15–17 kHz. Human hearing is sensitive to pitches within 2–5 kHz. In comparison, the cat's hearing is most sensitive to sounds within 500 Hz to 32 kHz.

A cat's hearing is so sensitive that sudden noises and loud voices can be extremely startling and normal human conversation is often too loud. This is why cats tend to feel more comfortable with people who walk and talk softly and speak in higher pitches.

There is a fascinating area of research called bioacoustics which is the study of sound in animals. Like humans, nonhuman animals are affected by music. In a study published by the *Journal of Feline Surgery and Medicine*, cats were played music while under anesthesia. Stress was measured by an increase in pupil diameter and respiratory rate. Heavy metal and pop music increased pupil diameter and respiratory rates, while classical music decreased them. Heavy metal caused cats the most stress.

This is not only important for those who live with cats but it's especially important for those who work with them in kennels, veterinary, or shelter settings. Although we might like rock music to keep us going, this isn't appropriate for cats. Playing light classical music, strings, or a soothing instrumental is a better choice for cats.

SCENT

A cat's sense of smell is about 20 times better than ours. They can smell and evaluate things from odors that we cannot. Cats, along with many other animals, have an olfactory organ called the vomeronasal or Jacobson's organ. This organ is located behind the upper incisors and is composed of two little canals or ducts that connect the mouth to the nose. You may have seen your cat sniff an object, or maybe even you, open his mouth and make a funny face or grimace. This is called the flehmen response. It is thought that the vomeronasal organ is used to 'taste' odors and pheromones. We don't have a similar organ to compare it to.

Cats can be so sensitive to odor they become reactive to it. It is common for aggression to occur between cats when one cat returns from the vet. This can even occur between bonded cats who have lived together for years. The cat who stayed home can't seem to recognize his former housemate so fighting ensues. The odor of the clinic seems to override any visual recognition he has.

Pheromones are chemicals produced by an animal that physiologically or behaviorally affect others of the same species. Cats leave pheromones when they rub their head, face, cheeks, and parts of their body on animals and surfaces. Some of the cat's facial pheromones have been recreated or synthesized into a product called Feliway. The idea behind its use is that when cats are exposed to it, it calms them. Not all researchers agree on its efficacy. I personally find it useful and helpful for resolving feline urination problems.

SPRAYING

Spraying is a form of scent marking carried out by intact males. On occasion, unspayed females may spray as well. Spraying is a normal behavior that is controlled by hormones. When a cat urinates, he digs a hole or squats, and then afterwards, he scratches the ground or surrounding area to cover it. With spraying, an intact male will turn his back to a vertical surface such as tree or fence post, tippy toe on his hindlegs, quiver his tail, and spray a foul-smelling urine. This urine tends to hit a vertical surface and drips down. The cat does not cover the sprayed area afterwards.

The strong smell or odor of spray is due to a sulfur-containing amino acid, **felinine**. The amount of felinine in a cat's urine correlates with how much testosterone he or she has. Intact males have very high levels of felinine in their urine. Neutered males and females have much less.

Spraying is usually territorial. If there is any conflict over territory, the frequency of spraying between males increases. Commonly sprayed objects are usually prominently located around the boundaries and entries or exits of the territory. After 24 hours, the sprayed urine loses its effectiveness which causes the cat to spray or mark the area again.

For neutered males and cats in multi-cat homes, spraying is usually due to conflict between cats. The cat who sprays is often the victim and less secure in interactions. Spraying in neutered males can also be due to feline urinary tract disease and infection, so if your cat suddenly starts spraying, it's good to have him medically evaluated by a veterinarian.

Sometimes cats urinate vertically. They are not spraying or marking. These cats will squat initially, but if there is a vertical surface next to or adjacent to them, they will raise their hindquarters and urinate on the wall. They will usually then turn around and try to cover the urine afterwards.

ALLOGROOMING AND ALLORUBBING

Cats say hello to each other by greeting nose to nose. When a cat rubs the sides of her face, mouth, lips, and chin on objects or people, it's called 'chinning.' Allorubbing is when cats rub their forehead, cheeks, side of their body or flank, and tail against another cat. Allogrooming is when cats mutually groom each other. They may sleep together, nuzzle, and lick each other.

'Bunting' is a social behavior when a cat rubs her head or face on a person. Rubbing around a person's legs can be solicitous and deferential, for instance, if a cat wants food or to be let out.

Nose touching, chinning, allogrooming, allorubbing, and bunting on people are affectionate or affiliative behaviors.

This is one reason why physical affection and sleeping with your cat are so important. Cats have sebaceous glands on their lips, chin, top of their head, and along the tail. When you sleep with your cat, not only do you smell the same and have a communal scent, this is how cats bond with each other and show affection.

BONDED CATS

Cats who are affiliated are often referred to as 'bonded.' When cats are bonded it means they trust each other, enjoy each other's company, and feel safe together (Figure 3.1). Bonded cats stay in close proximity to each other. When separated, they can experience stress and unease. Once reunited, they usually greet each other nose to nose with tails up, chirping, and/or allorubbing.

Since cats are very selective about who they choose for companions, separating cats who are affiliated or bonded can be traumatic for them. Unfortunately, bonded cats are routinely separated on intake by shelters and rescue organizations. The rationale being that it's easier to adopt out cats individually or as 'singletons.' However, this thinking comes at a behavioral price. Shy and less secure animals copy the behavior of older and more confident ones. Fearful animals, once separated from a close companion, can be harder to socialize and will take longer to adapt to new situations and people. This can lead to failed adoptions and relinquishments.

It can take a lot of time for cats to accept unknown cats. Sometimes, they don't accept new cats at all. People may want two or more cats, but instead of choosing cats who are already attached to each other, they choose cats who are unrelated or don't know each other. This can go poorly with one or more cats being returned or rehomed.

When looking to get a cat, especially from a rescue or animal shelter, don't separate cats who are bonded to each other, even if one is extra cute or appealing. Instead, look for a cat who is housed individually or as a singleton. When inquiring upon animals for adoption, ask open-ended and specific questions. Ask if the cat came in with other animals or cats. Often, this can be found in a shelter or rescue organization's database.

Figure 3.1 Bonded cats trust each other and feel safe together. © Orxan Musayev.

Since cats are routinely separated upon intake, unless you specifically ask the question and someone searches for it, you may never know. The cat you are interested in may have a companion who is with a foster or in the shelter too.

If you work at an animal shelter or rescue, especially in a leadership position, make it a priority to keep bonded cats together. Educate adopters on the benefits. These include the following:

- They provide companionship for each other.
- A person can be away for longer periods without having to worry that the cat will be lonely.
- Bonded cats are less likely to have separation anxiety.
- Young and energetic cats will get more physical exercise and stimulation by playing with each other.
- Young male cats like to wrestle, so if they are affiliated boys they will share the same play style.
- Affiliated cats are far less likely to show play aggression toward people.
- Cats who are bonded to each other are less likely to exhibit attention-seeking behaviors such as knocking things off of counters and meowing at night.
- If you don't want cats to sleep with you in the bedroom, they have each other to snuggle with.
- Cats are happier when they have companions they like.

SLOW BLINKING

Cats do not blink rapidly the way we do. Cats blink slowly and intentionally to convey peace, affection, and to show they are non-confrontational. Blinking is reassuring and friendly. You may notice your cat slowly blink at you in bed or when you softly speak to her.

A cat who does not want conflict may look at another cat, look away, and then slowly blink in that cat's presence. Slow blinking is an excellent signal that people can give to fearful cats. If you relax your gaze, and with intention, softly blink at a cat, you may see the cat's eyes soften and she will blink back. By looking at a cat, slowly blinking at her, and then looking away, you are telling her that you are non-confrontational.

VOCALIZATION

The most common feline vocalization directed toward people is meowing. Meowing is less often heard during cat–cat interactions and seems to be specifically directed toward us. Since humans are verbally oriented, when a cat meows, it gets our attention.

Purring is heard in cats and other animals who are solitary hunters such as civets, mongooses, genets, and hyenas. Purring occurs primarily during contact with other individuals and can be a sign of pleasure, comfort, and reassurance. Kittens will purr while nursing. Sometimes cats will purr to self-sooth, which occasionally can occur when a cat is anxious, hesitant, or in pain.

Chirping is an affectionate signal cats make upon greetings and when they anticipate something good they are about to receive.

Growling or yowling is a sign of anxiety or aggression. Growling means the cat feels threatened and it is made by the cat to create distance. Yowling is when a cat's emotions escalate and she begins to panic or a serious altercation or fight may break out.

Chattering occurs when cats see small animals such as birds at a window. This usually occurs when a cat doesn't have access to the animal. It is likely a displacement behavior mimicking the kill bite that cats perform when subduing prey.

HISSING

Hissing in cats is a commonly misinterpreted behavior. Contrary to popular belief, hissing is not an aggressive behavior, nor is it generally exhibited by an aggressive cat. Hissing is a defensive gesture. It is almost always exhibited by a cat who feels victimized or threatened in some way. Often, it is a way to avoid a physical confrontation.

In cat–cat dynamics and inter-cat aggression, the cat who hisses can be the victim or the one to be chased and antagonized by other cats. Hissing is simply an emotional expression of discomfort, fear, or stress. A hissing cat feels threatened and insecure.

For many cases of inter-cat aggression, the hissing cat is often confused with being the aggressor. This can occur when the cat hisses at a dog or a family member, as well. A common misconception is that the cat who hisses is 'teasing' or 'taunting' the other individual.

In many situations of inter-cat conflict, the aggressor doesn't hiss, but instead stares. Prolonged direct staring, following, stalking, taking over areas the other cat may have used or claimed, and chasing are all common behaviors exhibited by aggressive cats.

Hissing is simply an expression of emotion; 'I'm upset,' 'I feel threatened,' 'I'm uncomfortable,' or 'I'm scared.' Whether your cat hisses at veterinary staff (and is unfortunately labeled the 'mean' kitty) or whether your cat hisses at newcomers or guests to the home, she is feeling defensive and vulnerable. If you try to 'correct' or punish your cat for hissing, you will make a bad or scary situation worse for your cat and only make her more upset.

Occasionally, there are cats who hiss to express their displeasure even when no threat is present. These cats will often hiss but not direct their hissing at any particular individual. A human equivalent might be swearing when you forget or drop your car keys. These cats are rather comical. They may hiss to express disgust or disappointment. An example would be when your cat, wanting to enter a room, hisses after you gently maneuver her away from the door. Or, when you bring your cat in from being outside in the yard, she walks away and hisses in protest.

Some cats hiss when they experience pain. For instance, a cat with arthritis may hiss after jumping down from a window perch or chair, and a cat with back pain may hiss when you pick her up to give her a hug.

Obviously, there are many things that cats may dislike or can upset them. By being aware that when your cat hisses, she is likely feeling afraid, threatened, or uncomfortable, it may change or modify the way you interpret her behavior and any inter-feline dynamics.

BODY LANGUAGE

It's helpful to understand how to read a cat's body language and to know the differences between happy and content cats and fearful, anxious, or aggressive ones.

EARS

Cats' ears are mobile. They can move their ears or pinnae toward and away from sound. The position of their ears can also indicate their emotional state.

A content, alert, or happy cat will have her ears upward.

Ears that are pulled back, down, or flattened indicate extreme fear. This may be accompanied by hissing or growling and even lunging. Fearful cats are not 'mean' or 'angry.' Given a choice, they would rather avoid conflict and retreat (Figure 3.2).

You may have noticed that your cat's ears twist or swivel. Although a cat's ears can twist in response to noise, they also twist according to her mood and emotions. You may see both ears on a cat swivel when she is irritated, indecisive, defensive, conflicted, aggressive, restless, pent up, or annoyed. Cats' ears will twist during play or mock fighting and when they grab onto a stuffed toy and kick it with their hindlegs. Frequently, when a cat's ears twist, her tail will flick or wag, too (Figure 3.3).

If your cat's ears are winged or what's frequently called 'airplane' ears, this is likely an infection such as yeast or ear mites (Figure 3.4). This is especially true if she periodically shakes her head. Ear infections can be very painful. If your cat's ears have black or brown goo in them, her ears are not 'dirty.' Her ears should be checked and/or treated. The inside of a healthy cat's ears should be relatively clear and pink.

Figure 3.2 Ears that are pulled back, down, or flattened indicate extreme fear. © Dr. Sheilah Robertson.

Figure 3.3 Cats' ears twist during play, mock fighting, and when they are defensive, annoyed, aggressive, irritated, or conflicted. © Creative Commons_Dreamstime.

Figure 3.4 If your cat has 'airplane' ears or shakes her head frequently, she likely has an infection or ear mites. © Valentina Degiorgis_FreeImages.

Figure 3.5 When cats are fearful, their pupils will dilate and become round. © Sandy Preister.

EYES

When cats are content, calm, or happy, their eyes will be almond-shaped, soft and relaxed. The pupil size and shape changes with emotion. When cats are fearful, their pupils will dilate and become round (Figure 3.5). If a cat is more agitated or aggressive, the pupils will constrict. (This is often referred to as slit pupils.) The pupil size and shape also changes according to how much light there is. In low light, pupils dilate, and in brighter light, pupils constrict.

Looking vs. Staring

All animals are sensitive to staring. As humans, when we talk to each other, we often bob our heads, glance away, and add verbal confirmations, 'yep,' 'uh huh,' and so on. When we do this, we interrupt or break eye contact with each other so it doesn't get uncomfortable. If you spoke to someone and they stared back at you without nodding, glancing away, or giving any verbal confirmations or feedback, it would become uncomfortable quickly. This is when we feel the need to glance away or we might feel like someone is staring at us. The difference between looking and staring at someone is 2.5–3 seconds.

We tend to notice when cats *intensely* stare at each other or focus on a laser pointer or a chipmunk, but we often miss *passive staring* between cats. Cats will intentionally use staring to displace and intimidate other cats. This can be a form of passive aggression. I call it the 'creepy' or 'creeper' stare. To us, the cat who stares may look innocent, but the other cat doesn't perceive it that way.

Cats don't blink frequently the way we do. When cats are being non-confrontational or friendly to each other, they will look at each other, but then glance away again. If one cat stares at another for longer than 3–4 seconds without breaking eye contact, you may see the other cat look uncomfortable and glance away. She may look fearful, hiss or growl, or try to leave the room. People often mistake this behavior as the 'starer' wanting to make friends with the other cat, but that may not be the starer's intention.

When we say hello to cats or lean in to pet them, we look at them for far longer than three seconds without breaking eye contact. Cats are cute. But, when we stare at cats, it can be intimidating to them, especially if they do not know our intentions. Staring at a cat or locking eyes with them for longer than three seconds can be seen as a challenge or threat.

TAILS

Tails on cats are very expressive. By watching your cat's tail, you can understand or interpret your cat's emotional state, how it changes, or how your cat may be affected by or interacts with other animals or people in the home.

Tail Up

When a cat is relaxed, happy, confident, or anticipating something good, her tail will be held vertically like a little flag. This is a signal that indicates she wants to interact or engage (Figure 3.6). Cats will raise their tails upon greeting people or other animals they like. You may see your cat's tail rise when you wake up in the morning and come home from work, before playtime, cuddles, or meals. If your cat runs to you with her tail held vertically, take it as a compliment. It is.

Tail Tucked

If your cat's tail is tucked tightly under her body, it indicates fear. It can also indicate fear due to pain (Figure 3.7).

Tail Wrapped

If your cat's tail is wrapped tightly *around* her body, she may be cold or uncomfortable, or feeling distant, isolated, or does not want to engage. Cats like warm, soft surfaces to lie on, so a cat's tail may be wrapped around her when she lies on a hard, cold surface such as a wood floor.

Tail in Line with Body

If your cat's tail is in line with her body or parallel to the floor, she is likely in the middle of a routine or transitioning from one place to another, such as walking into another room or heading to her water bowl or litter box.

Figure 3.6 Cats will hold their tails vertically when they want to interact and upon greetings. © Aleksandra Sapozhnikova.

Figure 3.7 This kitten's tail is tucked tightly under her body which indicates extreme fear. © Anna Krivitskaia.

Tail Twitching or Flicking

Just as twisted ears can indicate indecision or irritation, a twitching or flicking tail can mean the cat is hesitant, indecisive, uncomfortable, irritated, annoyed, or bothered. A cat's tail may flick when she's in pain. If a twitching tail turns to wagging, it indicates the cat is becoming increasingly restless or may be aggressive.

Wagging

Cats will swish or wag their tails while hunting and during play or before pouncing and jumping on another cat. Bored and restless cats and cats with high play drives but no specific outlets to direct it to swish their tails back and forth regularly. If you pet your cat and her tail starts to flick at the tip, she is not particularly sure about your attention or the manner in which you are touching her. If the flicking turns to wagging, it might be time to stop petting her or to change your approach because she will likely scratch or bite you. The tail can also wag or swish back and forth with a loss of eyesight or balance.

Puffed and Fluffed

If your cat's tail becomes fluffed, this is a physiological response, similar to when our hair stands up on the back of our neck. This usually occurs if a cat is surprised or spooked. It's also common in kittens and young cats when they become over-aroused and excited.

Fluffed Up and Arched Tail

The cat's tail is in this position when she is torn between being aggressive and running away (Figure 3.8). This tail is frequently seen in the Halloween cat pose. It can also be seen when two cats play wrestle and when kittens practice the defensive-aggressive posture or sideways hop.

BODY POSTURE

Content or Happy

A content or happy cat will have a relaxed body and tail. He will not be hunched or crouched, nor will his tail be tucked or wrapped tightly around him (Figure 3.9).

Bread Loaf

A cat in this position is huddled and has his feet and legs tucked underneath him. His tail is wrapped around him and he faces down and forward. He may seem like he's relaxed, but he doesn't look quite comfortable (Figure 3.10). The bread loaf position can be a sign that the cat is cold, in pain, or emotionally or physically uncomfortable. He may also be bored, isolated, or waiting for something. It's important to take note of when and where your cat shows this posture (Figure 3.11).

For instance, cats within the same home will position themselves relative to each other with some personal distance between them. If one cat is huddled or hunched

Figure 3.8 The puffed up arched tail is seen when a cat is torn between fighting and running away. © Yulia Zhemchugova.

Figure 3.9 A content and relaxed cat. © Kari Shea.

with his tail wrapped around his body in the loaf position, there may be disharmony between the cats or he might feel isolated or uncomfortable. If a cat is in the bread loaf position, he may also be cold or on a hard surface. The moment you place a heating pad or soft sherpa throw underneath him, he will sprawl out or curl up and relax.

Figure 3.10 The bread loaf position is frequently seen when the cat feels isolated, cold, uncomfortable, or is waiting for something. © Creative Commons_Dreamstime.

Figure 3.11 Cats will be crouched or huddled when they are not feeling well or are in pain. © Supasit Chantranon.

Freezing

A cat who freezes doesn't move and is generally quiet. The cat may remain still when he is anxious, uncertain, or afraid.

Figure 3.12 Fearful cats try to make themselves small, lean or cower away, and often, avoid eye contact. © Anna Krivitskaia.

Fearful

A cat who is fearful will be crouched, huddled, or hunched. His ears may be flattened or pulled back and his pupils will be dilated. His tail will be tucked underneath or wrapped tightly around him. Fearful cats try to make themselves small. They want to hide or disappear, and they will lean or cower away from what frightens them. Fearful cats often avoid eye contact as well and may do this in between hissing or lunging (Figure 3.12).

With inter-cat conflict, a frightened, defensive cat will crouch down or if an altercation is about to break out, roll onto his back. This can also be seen when two cats are playing or mock fighting.

Additional signs of fear will be hiding, inactivity, skittishness or startling easily, rapid breathing, a fast heart rate, scurrying or slinking away, remaining close to the ground, and lip-licking. Cats have sweat glands on the pads of their paws, so when a cat is stressed or fearful, he will sweat from his paws. He may also excessively shed.

Aggressive

When a cat is aggressive, he will generally make himself appear larger and will face or directly stare at the threat. His hind end will likely be elevated. Many cats can be fearful and aggressive at the same time. In these situations, they feel they cannot get away. Murray (Figure 3.13) is a cat who is fearful and aggressive to strangers or guests in the home. In this photograph, his ears are flattened and his pupils are dilated, but he is not getting small or leaning backward. Instead, he makes direct eye contact. He will hiss at people if they get too close to him. He may also follow and lunge at them if they look directly at him or stare at him.

Figure 3.13 Murray can be aggressive to strangers and guests in the home. © Karen Simmons.

Halloween Cat (Arched Body and Tail Down)

A cat in this position will have an arched back. His tail will likely be arched at the base, but held down. The cat is torn between fighting and fleeing. It can be seen in both aggression and defense. The cat will prefer to run away and avoid conflict, but will fight, if he is forced to.

The Cat Gets Puffy (Piloerection)

If you see a cat get 'puffy' or fluffed, this is piloerection. Piloerection is physiological response, similar to when our 'hair stands up at the back of our neck.' Cats will get puffy when they are spooked, startled, or fearful. Young cats, especially kittens, can get puffy when they are excited, ramped up, and over-aroused. You'll see kittens exhibit piloerection when they practice their sideways hop (or defensive-aggressive posture). Active, young cats can become 'fluffed' during and after play or when they race back and forth performing 'zoomies.'

4

EMOTIONAL BONDING AND YOUR RELATIONSHIP WITH YOUR CAT

OUR RELATIONSHIP & PREVENTING BEHAVIOR PROBLEMS BASICS

STIGMAS AND STEREOTYPES

In our society, despite their popularity, cats are still stigmatized and have negative stereotypes. Part of this is because cats are behaviorally misunderstood. Much of this stems from the Catholic church in medieval Europe. In Christianity, it was, and still is, believed that God gave man dominion (or 'domination') over animals. Animals are there to be used by man and controlled by man. Since cats were independent and couldn't be dominated or controlled, the Roman Catholic Church considered cats to be allies of the devil. Dogs, in contrast, were preferred by the church because they were loyal, obedient, and subservient.

In 1232, Pope Gregory IX publicly denounced cats in his papal bull *Vox In Rama*, claiming that cats represented evil, the underworld, and were associated with Satan. Black cats were particularly evil since the Church believed that Satan appeared to his followers as a black cat. All across Europe cats were tortured and massacred. This was encouraged by the Church.

Cats were said to harm babies and sneak into nurseries at night to scratch, bite, and kill them. Cats were associated with women, Jews, and heretics. Women in the Middle Ages had a very low stature and were considered sinful. Women were associated with Eve, who caused the fall of Adam, so women were ultimately blamed for the fall of mankind. Accordingly, cats were associated with women and women with cats. Jews were also linked to cats. It was believed that Jews, considered 'Christ Killers' at the time, turned themselves into cats to sneak into Christian homes to cast evil spells.

During the 14th and 15th centuries, there were witch trials all across Europe and cats were thrown off of towers, tortured, and burned at the stake because it was believed that they aided witches.

It wasn't until the 16th century with the Protestant Reformation and then the Age of Enlightenment in the 17th and 18th centuries that the cat's image began to change for the better. As the Church's power declined and there was more literacy and freedom of thought, cats began to be looked upon more favorably and accepted as companions. In the 19th century, Queen Victoria permanently changed the cat's image because she not only kept cats as pets but she also bred and showed them. The breeding and showing of cats became a popular activity especially among the upper class.

DOI: 10.1201/9781003351801-4

Sadly, the medieval European mindset and stigma attached to cats still haunts cats to this day. Black cats are associated with witches, Halloween, and bad luck. Many people still carry the superstition that cats should be prohibited from nurseries because they can harm or suffocate babies.

Cats continue to be associated with women, especially older women, and are thought to smell. (Cats do not have body odor. The 'smell' that people associate with cats is due to unclean litter boxes and stale or old cat food.) Many people continue to prefer dogs over cats because, unlike cats, dogs are subservient and deferential people-pleasers.

We have the popular expression 'crazy-cat lady' for people who like cats and words such as 'scaredy-cat,' mean, nasty, fractious, and evil when describing cat behavior. We tend not to have these negative stereotypes and expressions when describing dogs or people who like dogs.

INDIVIDUAL VS. GROUP BEHAVIOR

Cats are behaviorally misunderstood. One of the reasons for this is that domestic cats are not group creatures. All group animals, including people, look to each other for guidance, direction, and approval. We follow the 'group' or what the majority of individuals are doing. Group (whatever we label it – herd, flock, school, or pack) behavior is influenced by and dependent upon the behavior and actions of others.

Cows, geese, dogs, people, fish, horses, sheep, and other group creatures mimic and copy each other. If one dog runs the fence, all the other dogs run the fence, or when one dogs barks, others join in. People behave the same way in groups. This is why people, cows, and sheep can be herded, but even the best wrangler can't herd or corral cats.

Group and communal animals are highly sensitive to what others in the group do and change their behavior accordingly. For instance, if a herd of cows are in a field, seemingly ignoring each other, when one notices something far off in the distance, all the other cows look in that direction. When we walk dogs, if we turn and look in a direction, the dog orients in that direction too. Often, we don't notice this, but we do notice that if we point to a dog, the dog looks at where we are pointing or motioning to. However, if you point to a cat, she will look at your finger, even if what you are pointing at is only a few inches away.

Cooperation and working together is necessary to keep groups intact. Geese fly in a specific formation and alternate who flies in front as they migrate. Hounding is when dogs are sent out to chase down or target an animal. A pack of dogs will work together to attack a fox or bear. If we picture a group of domestic cats trying to hunt or chase the same mouse, they will simply interfere with each other. A mouse is a meal for one and cats, for the most part, hunt alone or independently.

Because cats are not innately group animals, they are not seeking approval or disapproval from others, nor are they looking to others for how to feel and behave. Cats are unique individuals who make their own decisions. In other words, the cat is always right. Since cats have no need to work cooperatively with each other, there is no reason to appease, placate, or defer to anyone.

Ultimately, this means that cats have fewer strategies to navigate conflict or deal with stress. When a cat feels threatened or if there is conflict between cats, their initial response is to retreat or create and maintain distance from each other.

This does not mean that cats are antisocial or don't like companionship, but it does mean that cats need to feel safe and actually like those they are with. Being close to others, whether a person, cat, or another animal, is a voluntarily choice for cats and not an innate need for survival.

BONDING WITH A CAT

Cats are very selective and need to feel safe. A cat will become affectionate with you, seek your attention, and want your companionship, if he likes and trusts you. Cats avoid each other and maintain distance if they do not believe they are part of the same social group. This is why many cats can be quite loving and affectionate to those they know, but the moment an unknown visitor arrives, they run and hide. If a cat follows you, remains close to you, and seeks your attention, you can think of it to be a true form of flattery.

It is your job to bond with a cat. How affectionate we are with cats and how much we interact with them and care for them correlates with how much they interact with and care for us. Caring for a cat involves talking to him regularly, responding to his needs, soliciting him for play and interaction, cleaning his litter box, and making him feel safe. If you ignore your cat, he will not bond with you. He will likely distance himself or avoid you.

To establish a good relationship with a cat, verbally acknowledge him by saying his name and softly blink at him whenever he enters a room you are in and when you greet him. Speak softly and in sweet tones to cats. If you greet and acknowledge your cat regularly, he will bond more quickly to you and will likely solicit you for attention.

Let your cat sleep with you at night. Cats show affection to each other by nuzzling and grooming each other and sleeping together. Cats bond more closely with those they sleep with. In addition, indoor cats have less behavioral problems when they can sleep in the bedroom at night (Figure 4.1).

Figure 4.1 Cats bond more closely with those they sleep with. © Adam Kuylenstierna.

Do not punish or reprimand cats. Punishments such as yelling, stamping, or squirting your cat with water will either frighten your cat and damage your relationship or create further behavioral problems. Punishments inhibit the cat's ability to learn and increase his anxiety. Your cat will not associate your punishments with his behavior, but instead will likely associate them with you, the locations he's in, or with other individuals, whether people, dogs, or cats.

DO CATS PREFER WOMEN?

Many animals prefer women over men. One of the reasons for this is that women tend to walk lighter and speak more softly and in higher tones. Higher pitched sounds, with the exception of distress calls, tend to be associated with nurturing young, showing affection, and courtship. Lower pitched vocalizations tend to be associated with territorial aggression and adversarial behavior. This is true for most animals, including those who communicate above and below our hearing range.

Research has demonstrated that cats do prefer women and girls over men and boys. However, this is not due to their sex, but due to female behavior. Women are more likely to feed cats and get down to their level when interacting with them. Women stroke cats more frequently, and they speak gently and more often to cats at a distance (Mertens & Turner 1988; Turner, 1995).

HOW TO PLAY WITH A CAT

It's important to play with indoor cats. It encourages bonding between you and your cat and is mentally stimulating and fun for her. It prevents boredom and makes your cat happy. The witching hour for cats is very early in the morning, so if you have a young cat, she will have boundless energy and will likely be most active at dawn, dusk, and during the night.

For self-play, toys should be very small and light so that they can move a long way with little effort. Older cats usually have little interest in playing with toys by themselves and even young cats can get bored quickly, especially if toys are big, loud with bells, or hard and plastic. Good choices for smaller toys are glitter balls, fake mice, crinkle balls, spirals, hair ties, twisty ties, plastic bottle caps, and bouncing balls, ideally no larger than 2 inches. Scatter them throughout the house and rotate them regularly to keep them interesting. Offering new toys or rotating toys every few days, as well as changing the locations they are in, can create new play situations.

Large stuffed or plush toys are fun for cats to hold onto and 'bunny kick' (kick with their back legs) (Figure 4.2). Often these toys are stuffed with catnip. Ideally, these toys should be round, oblong, or rectangular and similar in size to a zucchini, cucumber, or softball and no smaller than a tennis or baseball. They should be left on the floor in a central and social location for your cat such as the living room floor next to a coffee table, near her scratching posts, or in the middle of the rug in the bedroom. If a stuffed, bunny-kicking toy is hidden in a corner or kicked to the periphery of the room, she won't use it.

Daily interactive play with poles, strings, and feathers is also important especially if your cat is the only cat or animal in the home, or if you have multiple cats who don't play with each other. Unfortunately, the way we play with cats can leave them frustrated and cat toys are primarily made for people, not for cats.

Figure 4.2 Large stuffed or plush toys are fun for cats to bite, hold onto, and 'bunny kick.' Sarafina © Paula Lichter.

Cats hunt small animals who are camouflaged such as mice, voles, chipmunks, grasshoppers, moths, crickets, sparrows, and finches. Attachments at the end of poles and store bought toys for cats tend to be hot pink, neon green, and brightly colored with feathers and bells or pompoms which are too large or loud to keep a cat's interest. We also tend to swing and dangle toys over the cat's head or in her direction, or we slide the toy away from the cat only to swing it toward her again. Prey rarely, if ever, chases, taunts, or follows the cat. Only young or truly boisterous cats enjoy tackling toys that fly at them.

Play should elicit your cat's hunting instinct. Although kittens will chase anything that moves, older cats fine tune their skills and want to hunt or stalk their prey. When cats hunt, they see a small animal, crouch down low and remain still (not wanting to be noticed), and wait for the right moment to pounce. The rewarding and stimulating part of playing for cats is watching and stalking the toy, which can be rather boring for us. Prey animals tend to be small, quiver, freeze, hop, slink, or scurry away and hide from the cat.

Rethink playing with your cat as a game of 'Hide the Critter' or 'Hide and Seek.' Most cats become entranced by a toy when it moves *away* from them. Move the toy or string away from your cat and make it slink, quiver, and hide around corners, under pillows and rugs, and behind boxes (Figure 4.3). This way your cat can stalk the toy, pounce on it, and then chase it again.

After your cat catches the toy a few times, end play with a handful treats. Place the food down next to the toy and then remove the toy. Feeding your cat after she catches the toy will give her a sense of accomplishment and signify that playtime is over.

If you have difficulty getting your cat to play or if she gets bored quickly, try some of these strategies. By being creative, you might get a tubby or lazy cat motived or a bored and resigned cat, possibly, even enthusiastic.

Keep interactive toys hidden until play time. After playing with your cat, don't leave the pole or wand on the floor or she will lose interest in it when you try to play with her again.

Figure 4.3 Prey always moves away and hides from the cat. © Lucie Hosova.

Cats often like the tip or end of the pole on wand toys more than the feathers or toys attached to them, especially if the poles are skinny and black. Slide and quiver the tip of the pole under pillows, blankets, and sweatshirts, and along the edges of rugs, sofas, or the bed.

If you have pole toys with large attachments at the end, remove the attachment so your cat can chase the string itself. Let your cat play with the elastic cord at the end of the pole and use the feather or attachment as a separate toy. Move the feather along the bed or floor, under rugs, and behind pillows.

Tie a string, ribbon, or long shoelace to your wrist or jeans as you clean, cook, and walk around your home. This is enjoyable for cats and easy for us.

When your cat is on the stairwell, move a toy or string along the sides or edges of the railing so it hides from your cat and then make it reappear again.

Poles on toys should be long, ideally 3 feet or more, so you can distance yourself from your cat as she's playing. If the pole is too short (12″ or less), you can become a distraction. In addition, the longer the pole, the easier it will be for you to move it behind and around obstacles, pillows, and furniture.

Get large cat tunnels. Cat tunnels are good for cats because they have multiple entries or openings so are fun for cats to hide and play in. Quiver or hide the toy along the edges and underneath the tunnel and move it along the rims of the openings. Your cat can attack it from multiple directions.

Create an obstacle course for your cat. Scatter pillows, bags, boxes, and tunnels on the floor. Sit in the center on an ottoman, chair, or coffee table and move the toy or string around and behind the objects you placed down.

Slide an 8″ × 11″ piece of paper or magazine along the floor. Many cats like to chase the paper and slide on it as you move it away from them.

Stack boxes together and create large holes for your cat to enter and hide in. Scatter treats or catnip within the maze of boxes, or move a toy or pole around the holes and openings.

Crumple paper into a ball and toss it on the floor. Let your cat beat up the paper or tear it apart.

Cut tiny pieces of paper and twist them into spirals so they are similar in shape to twisty ties. Because they are small and light, many cats enjoy batting them around.

Move a toy, string, or pole under bedding, blankets, and sheets and wait for your cat to pounce on it or chase it through the bedding.

Place tote bags on the floor and make the toy at the end of the pole hop into the bags and disappear.

Tie little pieces of felt into knots and scatter them throughout the home.

Take a large piece of fabric, sheet, or even a scarf and drag it around the house and on the floor, sofa, or bed. Many cats like to tackle, wrestle, and hang on to the fabric as you drag it around.

Place wrapping paper on the floor and scatter catnip on it. Move a feather or a pole under the paper.

Allow your cat to catch and hold onto toys. When your cat grabs or bites the toy, keep it still for a few seconds. When your cat lets go of the toy, move it again. If your cat can rarely catch the toy or string, he'll get frustrated and lose interest playing.

Before you go to bed, leave a bunch of novel items around the home for your cat to investigate. Place boxes and paper bags on the floor, put out a cat safe edible plant, scatter catnip on the rug or create a fort of blankets so your cat can be occupied in the wee hours of the morning.

PLAYING WITH DECLAWED CATS

I've had a number of clients with declawed cats. Many of these cats were adopted from shelters or obtained from someone who previously had the cat declawed. Although declawing cats is illegal in many countries, it is common practice in the United States. It's a myth that declawed cats don't end up at shelters or get rehomed. They are and frequently do.

Note: Declawing is a painful and invasive procedure. It is the amputation of the distal phalanges or toe bones. Mistakes in surgery are not uncommon. Pain and complications can last for months, if not years, and sometimes the entire lifetime of the cat. Declawing cats can create behavioral problems, such as general timidity, hiding and retreating, biting (especially associated with play aggression), and failure to use the litter box.

Declawed cats are at a disadvantage compared to cats with claws and need special accommodations when played with. Cats with claws can hold onto and grasp objects. Cats without claws walk on their cartilage. To compensate for their lack of claws, declawed cats awkwardly wrap their arms and paws around toys. When a declawed cat tries to hold onto and grasp a toy, it usually slips out of his paws. This can make play frustrating and difficult for him.

SPECIAL CONSIDERATIONS FOR PLAYING WITH DECLAWED CATS

(Cats with claws can have fun too.)

Although young cats may naturally want to fly in the air and leap up to 'catch' toys, avoid doing this with declawed cats. Instead, make play more about stalking and hunting. The majority of a cat's weight is supported by his forelegs especially upon landing after a jump. When cats are declawed, they land and put weight on their cartilage instead of bone. Declawed cats, inevitably, have complications with arthritis and back pain as they age.

Provide soft surfaces and rugs for declawed cats to walk and play on or play with your cat on a yoga mat. This is easy to do and fun since cats are naturally attracted to the spongy feeling of yoga mats. Running along tile or wood flooring is cumbersome and not good for a declawed cat's paws and forelegs.

Sometimes, attachments such as little mice and frogs made from natural fibers can be easier for a declawed cat to grip. Keep in mind, many cats won't like to play with a toy any larger than a ping pong ball. Large pieces of fabric or thick pieces of yarn can also be fun to play with. Smooth fabric, silky toys, and slippery strings can easily slide from a declawed cat's paws.

When your cat bites the toy, stop moving it. Once he lets go or paws at it, move it again. Since toys easily slip from the paws of declawed cats, we often don't realize that the cat actually caught the toy or string, so keep moving it. Keep the toy or string still for a few seconds once your cat makes contact with it.

Get large, stuffed cat toys for bunny-kicking. These are usually filled with catnip. Leave them in socially relevant and central locations for your cat. Since the toys are larger, your cat can hold onto them easier. Your cat can bite on them and kick them with his hindlegs.

Attach crinkled or crumpled paper to a string at the end of a pole. Crinkled paper is very light and easy for declawed cats to hold onto.

Crumple a piece of paper and toss it on the floor for your cat to bat around. Crumpled paper is light, easier to hold onto, and declawed cats can rip it apart with their teeth.

Soft tunnels are fun for cats and great for hiding because they have multiple openings. Slide or slink a toy or thick string along the edges, openings, and sides of the tunnel.

Slide a pole under and between the fibers of a shag rug so it burrows like a small rodent or slithers like a snake in the grass. Your cat will try to catch it as it moves through the carpet fibers.

Use the tip or end of the wand as a toy. Keep it flush to the floor as you move it *away* from your cat. Turn it into a game of hide and seek. Slide the wand under pillows, bedding, blankets, and rugs. After moving the pole, keep it still, but so it remains partially hidden. Your cat will pounce on the rug or pillows or dodge directly into the blankets after the wand.

Slide an 8″ × 11.5″ flat piece of paper along the floor or a rug. Cats will often slide along with the paper after they chase and pounce on it. Claws aren't needed for this game.

Put a bunch of wrapping paper on the floor and scatter catnip on it. Move a pole or toy under and around the paper.

To end play, give your cat a pile of treats after he catches the toy. This will give him a sense of accomplishment and mentally transition him to another activity.

CAT ALLERGIES

Many people have cat allergies or think they are allergic to cats. Those who think they may be allergic or who were told they were when growing up should get tested. The allergy may be mild, and sometimes, there is no allergy at all.

An allergy to cats means that there is an allergy to dander. Dander comes from the cat's saliva, not from the cat's fur, itself.

There is a strong psychological association with cat allergies. People who have had or think they've had an allergic reaction to cats can feel symptoms when they become aware of something they mentally pair with cats, even if no cats or dander are present. In other words, they can release histamine and have an allergic response to something harmless or benign, if they associate it with cats. For instance, a person might be happily in your home eating dinner until they see a cat scratcher or until somebody mentions there is a cat in the home. Upon seeing the scratcher or realizing there is a cat, they begin to have an allergic reaction. They may have been in the home, previously, for hours, with no prior symptoms.

When an individual dislikes cats or has a negative association with them, it is unlikely they will be able to improve or get over their allergy. However, if you are allergic to cats, but love them, or if you are involved with someone who loves cats and has them, there are things you can do to make life easier. Sometimes, it takes a bit of adjustment, but for many individuals, even with severe allergies, over time, they work. Some people can eliminate their cat allergy entirely.

APPROACH

Take medication. It's safe and effective. Your symptoms will be much milder to non-existent.

Wash your hands regularly, especially after petting cats, and try to avoid touching your face after you pet them.

Use a lint roller. When you are with cats, intermittently and after you leave the room or home, use it to remove excess cat hair that may be present. If you don't have a lint roller or don't want to use one, wipe your clothes with a damp paper towel.

Pair good things with the presence of cats – play with the cats, give treats to them, eat chocolate, drink lovely coffee, play your favorite music, or do something else you enjoy. If you have or develop a positive association with cats, you will likely have less of an allergic reaction.

IN THE HOME

Get an air purifier.

Place decorative throws over the bed, sofas, and furniture for the cat to lie on. When a person with the allergy visits, these throws or blankets can be removed and placed in a chest, bureau, closet, or washed.

Use a lint roller or damp cloth or paper towel to wipe down furniture to remove any cat hair and dander. Depending on the fabric, wiping down furniture with a rubber glove can work too.

Stroke and pet the cat regularly with a moistened or damp paper towel. Dander comes from a cat's saliva, not the hair itself, so by wiping down your cat you can remove excess dander the cat has on his fur.

Open windows. By getting air to circulate, it can mitigate the allergic response so the body can acclimate and adjust over time.

If possible, avoid rugs, especially that accumulate pet hair. Depending on the number of cats you have and the person's sensitivity or allergy, lightly damp mop wood floors on a regular basis.

Keep litter boxes clean. Wash and clean cat food and water bowls daily.

Vacuum.

FOR THOSE WITH AN ALLERGY

Hang out with the cat in increments. Wash your hands after touching or petting the cat. Pair the presence of cats with pleasant things such as nice food or desserts, a good movie, coffee, or your favorite music so you can develop a positive association with them. For people who love animals and cats, the presence of the cat alone may be positive enough.

If you have no cats at home or aren't used to animals or animal fur, place a very small amount of cat hair in your car or in your home. This allows you to adapt and be exposed to the allergen without overloading your body to cause a reaction. The goal is to expose yourself, gradually, to cats or to an individual cat at a level and in ways that don't trigger an allergic response.

If you decide to live in a home with a cat and the above recommendations don't help mitigate your allergic reaction, designate a cat free room. As you adapt and have less of an allergic response, you can gradually allow cats in that room.

MOVING

It can be very stressful moving with animals, especially long distances. Staying at hotels and traveling can be exhausting, but doing it with cats is even more challenging. Most cats hate car rides. They need to eliminate but can't go outside at a rest stop as dogs can. Some cats immediately vomit or have diarrhea in the carrier or crate within the first 20 minutes of driving or they wail and cry continuously as you apologize to them profusely.

Cats tend to be fearful in new environments, hate loud noises, and can be afraid of new people. Trucks passing on the thruway, stopping for gas, noise from hotel lobbies, and the possibility of escape can make traveling with cats physically and mentally harrowing. However, there are things you can do for you and your cats to make travel easier and more tolerable.

When doing local moves, ideally, set up a room in the new home with all your cat's belongings. Either move the cats into this room first, before you move in your furniture and boxes, or keep your cats in a room in the old home, with all of their belongings, while you pack up and move to the new environment.

Set up the new, safe room for the cats before you move them. Make it comfortable and inviting, with everything they need. Immediately, bring your cats into this room upon moving them. If the cats are frightened or nervous, once things have settled, let them come out to explore at night. If they want to explore earlier, let them to do so, but let them come out at their pace. Sleep with them in the room at night until they feel comfortable enough to be in other rooms of the home.

Do not let cats outdoors, unsupervised, after any move because they will not be confident enough in new surroundings or have their bearings.

FOR CAR TRAVEL

If your cats get stressed in the car, ask your veterinarian for a behavioral sedative for car rides.

Microchip your cats before relocating and make sure the addresses and contact information are kept up to date.

Get tracking collars. Keep these on your cats at all times until you get to your final destination. Not only will this give you piece of mind, but if something happens, you will immediately know where they are.

Call ahead of time to make reservations at hotels that take cats. La Quinta's are pet-friendly and AAA has an online listing of hotels that accept animals. If you make reservations ahead of time, you'll have no issues and have a safe destination to go to after hours of driving.

Always bring a double set of car keys.

When traveling with cats, use large soft dog crates or carriers. Cats can feel safe and hidden, but will be comfy and have room to stand up and turn around. You may even have additional space for a litter box.

Line the bottom of crates and carriers with pee pads, either housetraining pads for dogs or bed pads for people (often called chux pads). This way you can easily clean up if there are spills or if a cat urinates or vomits. Likewise, because these pads are absorbent, your cat won't have to sit in wet urine or liquid if he has to vomit or eliminate while you are driving.

Have everything ready. Paper towels, handwipes for yourself, unscented baby wipes for your cat, enzyme cleaner to pick up any accidents, and garbage bags so you can dispose of anything quickly when you need to. Bring extra water, water bowls, plastic or reusable forks, paper bowls, and paper plates. The last thing you want while traveling with animals on the road is to be unprepared.

If you put your cat in the back area of a car or SUV, make sure there is padding. We take for granted that the seating in vehicles is heavily cushioned for us. The moment you pull back the seats all there remains is metal and plastic above the tires on the road. This will make your cat nauseous and fearful. I don't recommend putting any animals in the far back of the car since rear end collisions are extremely common, and when an animal is placed in the back, they will be the first to get hit. It's also a bumpier ride (think of the back of a school bus), so the ride can be extremely unpleasant and it increases the likelihood of motion sickness. This applies for dogs, too.

Stopping at rest areas can be stressful especially when you have to leave animals in the car. If you are driving solo and don't have the option of taking turns with a partner, park in the front of the building where there are cameras. Lock the doors and keep your blinkers on. This lets people know you are immediately coming back and makes people less inclined to break in or do something they shouldn't. In warm weather, when you make a stop, keep the AC on. By having a double set of keys, you can keep the car running from the inside, but lock it from the outside.

Depending on the length of the drive, cats may need breaks. They have to stretch their limbs, eliminate, and eat or drink, yet they need to be and feel safe. When at rest stops, lock the doors and let them meander for a while. Give them snacks and a little food. Let them use the litter box and provide them with water. Once everyone is secure and settled, you can start driving again. Don't open car doors or windows until you know everyone is safe and accounted for.

Once you get to your destination, set up the cats in their safe room with all their belongings. If you are going to be rearranging furniture and moving boxes, which means doors will be opening and closing regularly, or if any of your cats are anxious, keep them in the safe room until you are set up. You can then let them out to explore in the evening or when they are comfortable.

REFERENCES

Mertens, C. & Turner, D. C. (1988). Experimental analysis of human-cat interactions during first encounters. *Anthrozoös*, 2, 83–97.

Turner, D. C. (1995a). *Die Mensch-Katze-Beziehung. Ethologische und psychologische Aspekte*. Jena and Stuttgart: Gustav Fischer Verlag/enke verlag.

5

HUMANE HANDLING

INTRODUCTION

Many of us regularly handle cats. We have to give them medications, clean their ears, brush or groom them, wipe their eyes, take them to the veterinarian, put them into carriers, cut matts, clean or wipe soiled bottoms and tails, remove ticks, clip nails, and give sub-q fluids and injections to older cats or cats with medical issues. We may have to pick up and carry cats even when they don't like being held.

Those who work at shelters, especially at intake (when people drop off animals), are required to take already scared and traumatized cats out of carriers and transfer them into cages or other carriers. (People who drop off cats usually want their carriers back.) The cats are then vaccinated, dewormed, weighed, and their ears are swabbed or sampled in loud, chaotic environments. These cats might have their blood drawn, as well. In the US, many states don't require formal training or licensing for veterinary technicians and many animal shelters don't have them.

When people trap or rescue cats, they move them from traps into carriers, or from carriers into new environments, transport them to veterinary clinics, give cats topical, oral, and injectable medications, bathe and groom them if they are dirty or matted, approach cats in cages, pick up and carry cats, socialize feral kittens, bottle-feed kittens, and get fearful cats used to touch and handling.

If we care about cats, we want to interact with them in a safe and positive way. We also want them to be receptive to being handled in the future. Many popular handling techniques are harmful to cats, have damaging behavioral side-effects or fallout (sometimes, that can last the lifetime of the cat), and increase the risk of injury to both ourselves and cats.

Although rough handling might seem momentarily effective, it has grave consequences. When one person roughly takes a cat out of a carrier or picks up a panicked cat by the scruff, it might work temporarily, but then she passes that cat onto others. She sets up the cat, and any people who then have to handle or interact with the cat, for misery and failure, and the very real potential of getting hurt. Humane handling techniques are kinder, safer, effective, and ultimately faster than rough handling or grabbing, pinning, and scruffing cats.

DOI: 10.1201/9781003351801-5

WHY OUR ATTITUDE MATTERS

First, we must have a positive attitude. If the cat hisses, growls, or swats at you, don't call the cat nasty, evil, or mean. These cats are simply frightened and fearful.

When communicating a cat's behavior to others, try to do so in an objective and informative way. Instead of negatively referring to the cat, describe what the cat is doing. State what the cat's behavior is. For instance, instead of saying the cat is 'mean' or 'nasty,' you might say the cat 'hisses when I reach for him' or 'the cat growls when I approach and will not come out of the carrier.'

It's okay to use words like afraid, frightened, timid, or avoidant. Words like hissing, hiding, lunging, will not come out of the carrier, or swatting explain what the cat is doing but does not negatively label or malign her.

If someone tells me they need help because the cat is 'mean' or 'nasty,' I don't have any clarity on what the cat needs or what I'm supposed to do. If I'm told that the cat growls on approach and won't come out of the carrier, I know I need to work on soothing the cat and finding a way to examine her while she stays in the carrier or a way to remove her from the carrier.

ALTERNATIVES TO SCRUFFING CATS

Many of us scruff cats. 'Scruffing' is when we grab the cat by the back of the neck to move or carry them. This immobilizes cats which is why it has traditionally been used. We learned that scruffing cats is the way the mother cat carries her kittens so it's natural for them to hang this way or it's what we've observed others do – at the groomer, vet, shelter, and even at home.

Mother cats have no choice but to carry kittens in their mouth and they do this when their kittens are newborns. Mother cats are very sensitive to how tightly they hold their kittens. As kittens get older, even after a few weeks, the mother no longer carries them this way.

Although scruffing *may* work temporarily, since the cat is powerless to move, she will become even more frightened of being handled in the future. Scruffing *increases* the likelihood the cat will bite and increases the risk of injury to both the person and the cat (Figure 5.1). Scruffing and dangling the cat by her neck can cause serious eye damage like glaucoma and cataracts, since it increases intraocular pressure. It is especially harmful for cats with uveitis or other eye issues.

Many cats are easily handled when people are gentle and use the right approaches, but the moment they are even slightly scruffed, they hiss, growl, flinch, cry, panic, or try to get away. Scruffing intensifies any fear and anxiety the cat may already be feeling.

Fortunately, there are other ways to handle, pick up, and hold cats. We can use a gentle approach, along with humane handling techniques, to control cats without scruffing them. This makes it easier for others to then touch, handle, or pick up the cats and easier for us to handle them again without further traumatizing them.

For instance, instead of hanging a cat by the scruff, we can rest our hand on the cat's head or the back of her neck, so she remains still. Carrying cats by supporting them in a football hold can prevent them from wiggling or squirming, see p. 64. Towels can be used to control a cat's movement while keeping her calm. Placing cats on

Figure 5.1 Although scruffing temporarily immobilizes the cat, it makes handling the cat in the future more difficult. © Oleg Troino.

a soft, non-slip surface can prevent them from losing their balance or feeling unstable so will prevent any escalation in aggression or reactivity. We can distract and redirect cats, especially kittens and young cats, with food and play, during and after handling, to make it less stressful for them and easier for us. If a cat has any abscesses or wounds, instead of alcohol or an astringent, we can use Betadine (10% Povidone-Iodine solution) or Neosporin cream which don't sting. If a cat is afraid in a carrier, we can remove the cover and examine her under a towel while she stays in the bottom half of the carrier.

APPROACH: BE QUIET, SPEAK SOFTLY, AND WATCH YOUR BODY LANGUAGE

Although the 'get it done' mentality and handling cats roughly is often seen as faster, it's not skillful or humane. Cats don't know how to handle aggression, so it's vital to go slow with cats and stay, as best as you possibly can, within their comfort zone. If a cat starts to panic, it's important to let her calm down before you continue or wait for another time.

If you work in a shelter and are in charge of staff or volunteers, designate the gentlest and most skillful handlers to be the ones who teach others. If you volunteer or work in a shelter or an environment where multiple people handle cats, learn from those who are the gentlest and calmest around cats.

How we approach cats can greatly affect their behavior. It's important to approach cats from behind them or from the side when you handle them. Try not to tower over them or face them directly (Figure 5.2). This is especially true for fearful cats. Frightened cats will prefer to face away from you when being handled. Avoid direct eye

Figure 5.2 Approach cats from the side or slightly behind them. © Oleksandra Polishchuk.

contact which can be scary, creepy, and startling to cats. Instead of staring at the cat or locking eyes with her, look at the cat softly, but then avert your gaze.

Cats have extremely sensitive hearing, so speak softly and quietly around cats. Loud and sudden noises are especially scary. Be cognizant of this when opening and closing carriers, doors, and cages. Instead of swinging doors open or letting them slam behind you, open and close doors thoughtfully.

Avoid rapid, sudden gestures or quick movements around cats. Instead, be calm and move slowly, but steadily.

If a cat becomes too scared or begins to be reactive, give her a few minutes to calm down before continuing. The more frightened or reactive she gets, the harder it will be to handle her. Don't yell at the cat if she squirms or say 'No' if she resists, and don't repeatedly tap the cat on the head as a distraction. These approaches will only frighten her more.

Cats will be calmer and more cooperative if you go at their pace and stay at their comfort level.

TOWELS

The use of towels and fleece blankets can aid in gentle handling. Towels can be used for comfort and to control the cat's movements. Towels also prevent injury. Towels and small blankets should be thin enough so that you can feel the cat's body and the position she's in, yet thick enough to prevent injury.

Do *not* use towels to wrestle, chase, or tackle the cat. If towels are paired with rough handling and scary procedures, cats will then become fearful and reactive or try to hide when they see a towel.

When fearful cats want to hide, they can panic if prevented from doing so. By lightly covering the cat's head while you examine or handle her, a fearful or timid cat will remain still (Figure 5.3). Some cats are calmer when they are carried in a towel or when a soft blanket is gently wrapped around them.

Figure 5.3 Gently covering a cat's head with a towel can make a fearful or timid cat remain still. © Dr. Sally Foote.

TOUCH, HANDLING, AND RESTRAINT

Try to think of everything you need, or may need, before handling a cat. Although it seems common sense, we often don't prepare fully. Prepare in advance. You don't want to be in the middle of handling your cat and then realize you forgot something you needed or can't find the right items.

Cats prefer light and steady touch over rapid petting or repeated stroking along the back. If a cat does not like how you touch him, he may show some of the following signs: crouching or cowering, flinching, pulling away, squirming, flattened ears, twitching or wagging tail, and crying or vocalizing.

Cats are often more comfortable when you gently massage their head and chin from behind them while you look at or examine them (Figure 5.4). For the most part, cats prefer to be touched and massaged on the face, head, and neck, specifically, on top of the head, forehead, and occiput – the little bony protrusion at the base of the skull (sometimes called the 'knowledge knot'), on and around the cheeks, and under the chin. Cats dislike their lumbar region (hips and waist area), lower legs, and paws touched. Of course, any area that is painful will cause the cat to retreat or act out if touched there.

Let cats choose what position they prefer to be in. If they prefer to sit while you handle them, let them sit. If they prefer to lie down, let them lie down. If they prefer to stand, do as much as you can with them in a standing position.

Depending on the setting, try to handle the cat where he is most comfortable. This might be on your lap, the floor, on a table, bed, or in a carrier.

Always place something soft underneath the cat such as a towel or cut yoga mat (Figure 5.5). Do not handle cats on slippery surfaces, laminate counters, or hard, metal tables. These surfaces are cold and provide no traction or grip for the cat which is why they struggle. In addition, metal tables are very loud which further frightens them.

When handling cats, it's important to fully support their body. This means their belly and torso should be stable and supported or all four paws should be on a steady surface (Figure 5.6). When a cat is tipped or off-balance, he will try to right himself. This makes people struggle with the cat and restrain harder, which makes the cat more reactive.

Figure 5.4 Cats prefer to be touched and massaged on their face, head, and neck, under the chin, and around their cheeks. © Huỳnh Tấn Hậu.

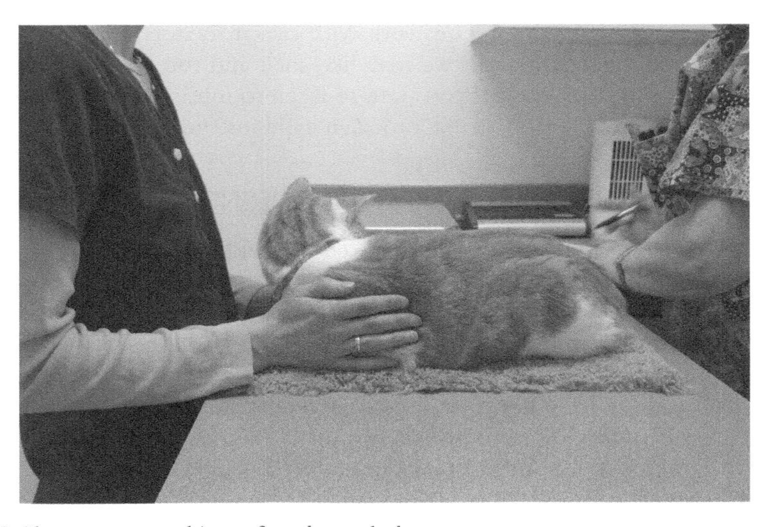

Figure 5.5 Always put something soft underneath the cat.

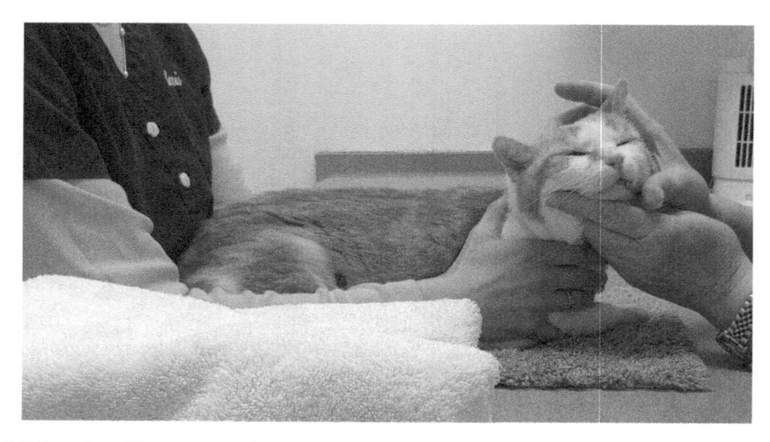

Figure 5.6 When handling cats, make sure their body is fully supported.

When holding and moving the cat's limbs, keep the limbs in their natural range of motion. It's better to move a limb from above and while supporting the joint rather than to pull on legs and body parts. Moving limbs and body parts the way they easily move or in their natural alignment is called passive range of motion. Often, this in itself can make handling animals much easier.

For instance, our wrist naturally moves up and down, but if you try to move it left and right, there is only so far it will go. If someone gently pushes forward on the back of your arm or triceps area, your arm will naturally extend with little effort and no discomfort. However, if someone tried to pull your forearm, you would be uncomfortable, resist, and pull back. A little twist or tweak on a limb or joint in the wrong direction is painful, so we naturally avoid it and pull away, as do cats.

WHAT ARE YOU LISTENING TO?

Music affects cats. Light, classical music relaxes cats, while rock and heavy metal stresses them. This is important for those who work with cats at shelters, veterinary practices, and when housing them. Although we may like rock and roll to keep us going, this music for cats adds to their stress. If you want to listen to music while working with or handling cats, play light classical, string, or a Zen-type instrumental.

HOW TO PICK UP AND CARRY A CAT (WHO DOESN'T LIKE IT)

Many cats are uncomfortable with being picked up, held, or carried. We tend to pick up cats from the front and cradle their back legs or we tip them vertically and rest their belly or chest against our chest and shoulders. We also tend to turn cats on their back to cradle them as if they were a human baby. These ways of holding cats can make them feel trapped and off-balance. In addition, cats can push off with their legs against our chest and arms causing us to get injured or scratched (Figure 5.7).

If you have a cat who dislikes being picked up or held or squirms and wiggles out of your arms when you attempt to carry him, try this approach. First, when picking

Figure 5.7 We tend to pick up cats from the front, tip them, or hold them vertically with their belly or chest against ours. © Maxim Mushnikov.

Figure 5.8 Football hold.

up your cat, don't face him. Instead, position yourself slightly next to or behind him. Place your arm over his back with your forearm under his belly and chest to support his torso. Your opposite hand can be placed on his chest to control his front legs (which I prefer) or it can gently rest on the back of his head and neck. Keep his spine straight and his body aligned. Allow his back legs to dangle. His body should be supported by your arm and body.

Hold him similarly to how you would carry a football (Figure 5.8). Keep his body parallel to the ground and his face oriented away from you and in the direction you are bringing him. Try not to tip him vertically.

When placing him down, be steady and make sure all four paws are on the surface or ground you place him on before you release your hands from his sides. Don't let him jump from your arms or drop him. Give him a handful of treats after you place him down, but while he's still positioned away from you.

Practice lifting and carrying him short distances. Initially, pair every pick-up with food. When you feel a little more confident, pick him up and carry him short distances to places he wants to go such as on the sofa or bed, the kitchen sink to drink water, or a windowsill to look at birds out the window. By picking him up and carrying or 'teleporting' him to where he wants to go and pairing your touch and pickups with food or something else he enjoys such as catnip or a sunny location, he will begin to have a positive association with being picked up. In addition, holding and carrying him this way is more comfortable for him and prevents injury to you and the cat.

Once he gets comfortable with being picked up, held, and carried, hold him for a little bit longer, carry him further distances, or begin cradling him slightly. Try to place him down before he feels the need to squirm away.

PICKING UP AND CARRYING A FAT CAT

Some cats are too large to comfortably carry in a standard football hold. In that case, it may be easier resting them against your stomach and chest, keeping them parallel to the ground and their body perpendicularly to you, as if you were carrying a medium- to larger-sized dog. Alternatively, wrapping a blanket around the cat and carrying him in a burrito wrap can also support him while keeping you safe.

GETTING YOUR CAT ACCLIMATED TO A CARRIER

It's common for cat parents to struggle to get their cats into carriers. The usual scenario is the person gets the carrier, stored in a closet, basement, or spare room, and upon seeing or hearing it, the kitty runs away terrified and hides under the bed or in a closet. Sometimes, a battle ensues. People try to grab and hold onto the cat while the cat claws, scratches, and bites to get away. Many times, cat owners call the vet to cancel appointments. Other times, people avoid vets or travel altogether, unless in a dire emergency.

If your cat is terrified of the carrier and runs and hides at the sound or sight of it, ditch it and get a new one. Most cat carriers are too small for adult cats. Start fresh with a medium- or larger-sized soft dog crate or plastic cabin crate. The larger the crate or carrier, the more room your cat will have and the less trapped she will feel. Cats are approximately 12″ tall and 18″ long, not including their tails, so the entrance and carrier itself should be tall enough for your cat to walk into and stand up in. If the carrier is too small, it will be difficult to place your cat into it without a struggle.

I personally use soft dog crates for my cats for quick transport (with carry straps) and large soft dog crates for travel.

If you have a cat who might panic in the carrier or scratch at the sides of it, a soft crate may not be the best option. This isn't a problem for most cats, but if this is your cat or you're trapping and transporting cats for rescue, a large plastic crate where the top can easily be removed may be better.

Acclimate your cat to the carrier in a positive way. Don't leave it in the closet or storage room only to bring it out for car trips and vet visits. Instead, leave it in the living room, bedroom, on a dresser, underneath a window, or in another preferred location for your cat.

Place a soft plush blanket, thick sherpa fleece throw, or fluffy bed in the carrier. Make the carrier comfy and accessible. Cats love warm, soft, plush surfaces.

Leave the door to the carrier open. If using a soft crate, roll up the mesh or canvas openings so your cat can easily enter and exit at will. For soft crates, once your cat is comfortable, you can release the mesh or fabric door, but leave it unzippered.

Place highly palatable treats, goodies, and fresh catnip in the carrier. If your cat is playful, play with her in and around the carrier. Whenever your cat goes into the carrier, play with her, praise her, and give her somethings she likes.

When it's time for a vet visit or road trip, calmly pick up your cat and place her in the carrier. (See 'How to Pick Up and Carry a Cat,' p. 64, and 'Putting a Cat into a Carrier,' p. 68) Try to remain calm while you zipper the entry or close the carrier door. When you return from the vet, place the carrier in a comfortable spot and let your cat out. Then place the carrier back in its original location.

By turning the carrier into a safe space and resting area, your cat won't be fearful of it. She will begin to view it as a safe place and area of comfort. Since the carrier is out all the time and she regularly uses it, she won't pair it with veterinary visits or car trips.

THE RIGHT CARRIER FOR THE CAT

When choosing a carrier for your cat, pick one that has multiple openings and a large entry so he can easily enter. The average sized cat is about 12″ tall and 18″ wide (not including the tail). Many carriers are too small for adult cats. Often, soft dog crates are better-sized and more appropriate. Do not try to push or squeeze a large cat into a small carrier. It will be very stressful for your cat as well as logistically cumbersome for you. The carrier in Figure 5.9a is too small for this cat – who at the time of this photo was only 10 pounds and underweight. The medium-sized crate (Figure 5.9b) is more appropriate. A good-sized carrier should be large enough for your cat to easily enter, stand up, and turn around in.

PUTTING A CAT INTO A CARRIER

Be sure the carrier and carrier entry are large enough for the cat, otherwise you will struggle.

Place the carrier on a countertop or table so it is at your waist level. Leave the carrier door open. Pick up the cat from next to or slightly behind him. (See 'How to Pick Up and Carry a Cat,' p. 64.) Place your arm over his back with your forearm under his belly and chest to support his torso. Your opposite hand can be placed on his chest to control his front legs or it can gently rest on the back of his head and neck. He should be parallel to the ground and facing away from you. Your arms and body should support the cat. Keep your body slightly behind him so he cannot back out as you put him into the carrier. You can also use a small kitchen-sized towel or bandana to cover his head as you place him into it (Figure 5.10a and b). This can prevent the cat from panicking. Once he is in the carrier, quietly and steadily close the door.

If your cat wiggles or squirms or you are nervous to pick him up, use a soft throw or towel. Position yourself next to and slightly behind him as you drape the

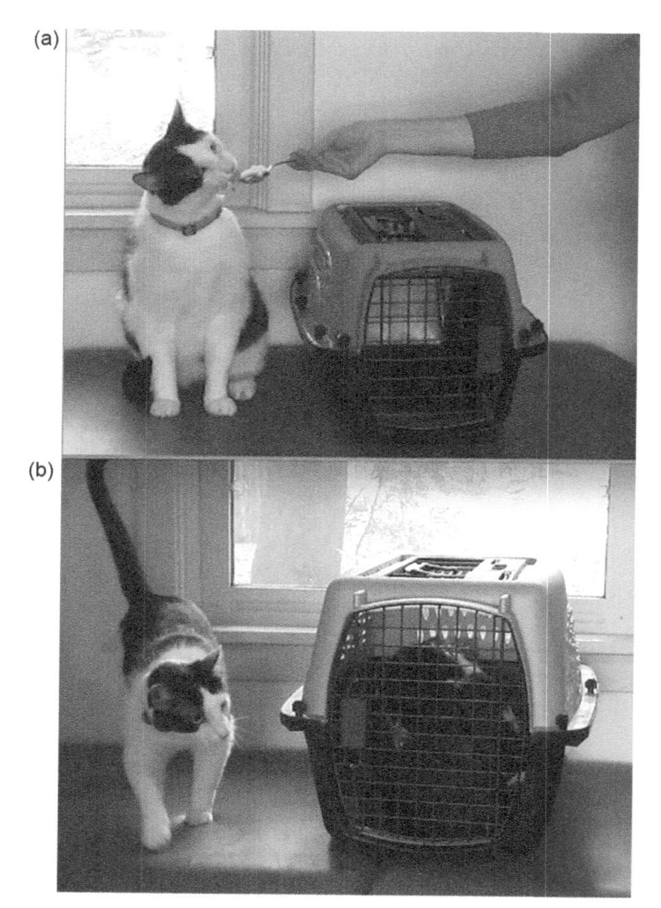

Figure 5.9 (a and b) The carrier (above) is too small for this cat, who at the time of this photo, was only 10 pounds. The medium-sized carrier (below) is more appropriate.

towel around him to pick him up. He should be positioned away from you and held parallel to the ground. His body should be supported by your arm and body. After you put him into the carrier, use the towel to block off the entrance so he doesn't back out.

If for some reason, you cannot pick up your cat, place the carrier in an area he frequently hangs out in. Put a warm blanket in the carrier and a few treats. See if he enters the carrier on his own. When he is in the carrier, close or zipper the door.

Alternatively, position the carrier on the bed or sofa adjacent to where your cat sleeps. Leave the entry open. When your cat is near or next to it, from behind him, lift or scootch him into it. You can use a soft bandana or towel to cover his head as you place him into it or a small towel to cradle him from behind so you can block the entrance in case he wants to back out. Remain calm while you close the carrier door.

Figure 5.10 (a and b) Placing a cat into a carrier.

REMOVING A CAT FROM A CARRIER

There are a few ways to remove a cat from a carrier depending on the carrier style, as well as the behavior of the cat. Although it's commonly done, don't tip the carrier upside down to dump the cat out. This approach will frighten her. Likewise, don't reach in and pull the cat out of the carrier by her scruff. A defensive and frightened cat is more likely to panic and be reactive. This increases the risk of injury to both you and the cat.

We often take cats out of carriers before we need to or earlier than necessary. With carriers that have multiple openings, many routine handling procedures and examinations can be done while the cat stays in the carrier. If the carrier comes apart, the cat can be examined and handled while she remains in the carrier bottom (Figure 5.11).

First, see if the cat will come out on her own. Speak softly and introduce yourself to the cat ('Hi Kitty'). Offer treats. Wait a minute or two. If the cat doesn't come out of the carrier and it comes apart, remove the top part of the carrier, and gently cover her with a soft towel. A towel can make a cat who wants to hide feel more comfortable and prevents escape. Do as much as you can with the cat under the towel while she remains in the bottom half of the carrier.

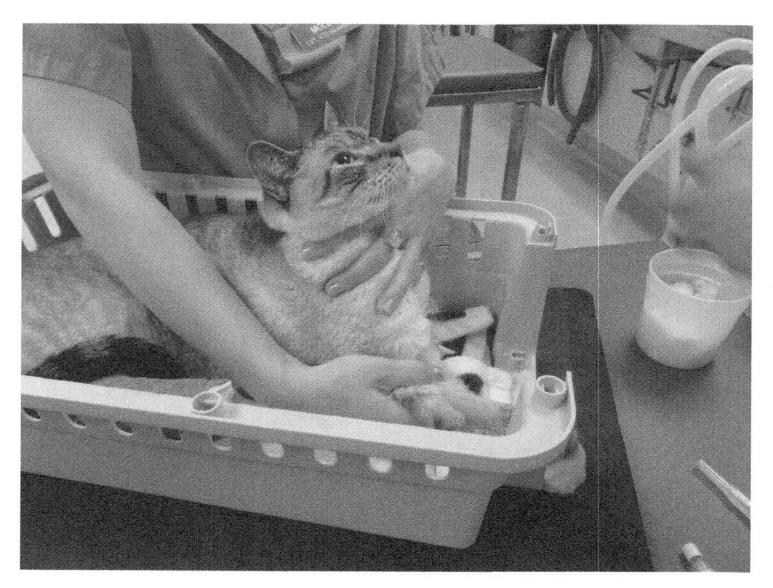

Figure 5.11 If the carrier comes apart, remove the top and examine and handle the cat while she stays in the carrier bottom. © Dr. Sheilah Robertson.

If there is no way to remove the top part of the carrier, but there are top or side openings, use a towel to cover the cat. Then tuck the towel around the cat's sides and gently lift or remove her. Make sure to fully support her with your body and the towel (Figure 5.12).

To weigh the cat, weigh her and the carrier together so you don't need to remove her from it. Then, once she's removed, weigh the carrier. If the top part of the carrier comes off, let her stay in the bottom part of carrier as you weigh her. Then weigh the top half of the carrier.

TRANSFERRING A CAT FROM ONE CARRIER TO ANOTHER (OR FROM A TRAP TO A CARRIER)

Especially at shelters, workers and volunteers frequently have to remove a stressed cat from one carrier (the carrier the cat was brought in) and move her to another one (a carrier belonging to the shelter or rescue). This can be very stressful for both people and cats.

To transfer a cat from one carrier to another, depending on the cat and situation, try one of two things.

Face the entry of the carrier the cat is in to the entry of an empty carrier. If there is a difference in the size of the carriers, use a towel to cover or block any empty space. If you wait a minute or two, the cat may move from one carrier to the other especially if the other carrier is larger and has something soft in it. Keep the towel positioned to block any gaps as you close the door to the new carrier (Figure 5.13).

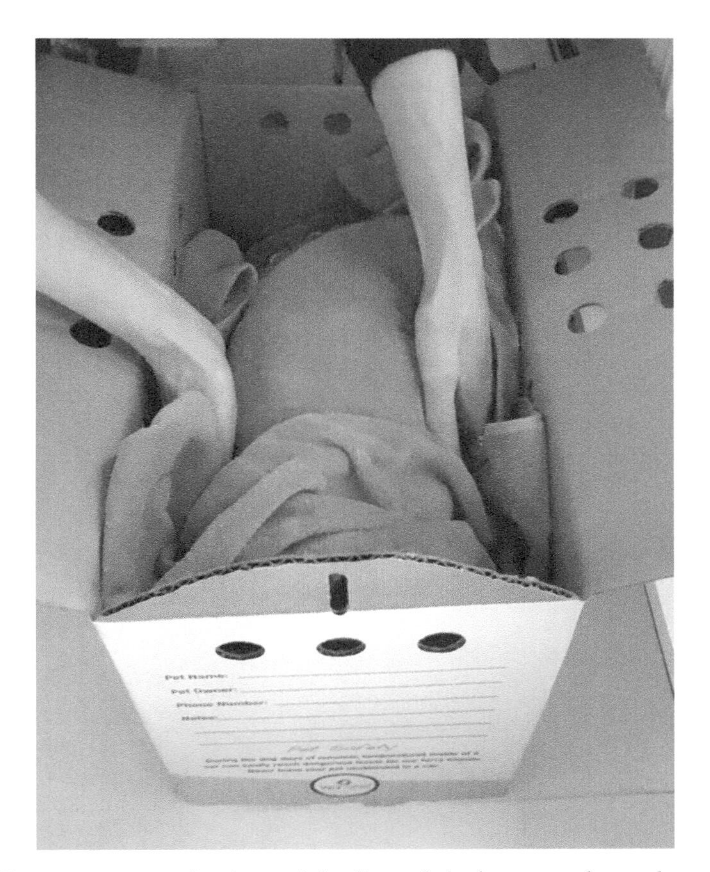

Figure 5.12 For a top entry carrier that can't be dismantled, place a towel over the cat and tuck it around her sides. Gently lift her supporting her body with your body and the towel.

Figure 5.13 Transferring a cat from one carrier to another.

Many cats after they've been trapped or in loud environments, especially in shelter settings, are far too scared to move. In this case, line up the carriers so the entries face each other. Carefully, remove the top part of the carrier the cat is in, as you gently slide a towel over the cat. This lets the cat continue to hide, prevents escape, and provides safety. When the top half of the carrier is removed and the towel is over the cat, 'scootch' the cat into the new carrier. Many cats will enter on their own. Once the cat is in the new carrier, carefully close the door as you use the towel to block off any gaps or openings (Figure 5.14a and b).

If the cat is in a box trap, line up the entry or opening of the trap to the entry of the carrier. Ensure the carrier is large enough for the cat. Place something soft in the carrier beforehand. Cover the opening to the trap and the entry to the carrier with a towel to block off any gaps or empty space. Often when you open the door to a trap, cats exit. Because the cat's going into another trapped space, use the towel or blanket to seal off the carrier entry after she enters so she doesn't back out. Slowly and steadily close the door to the carrier as you remove the trap.

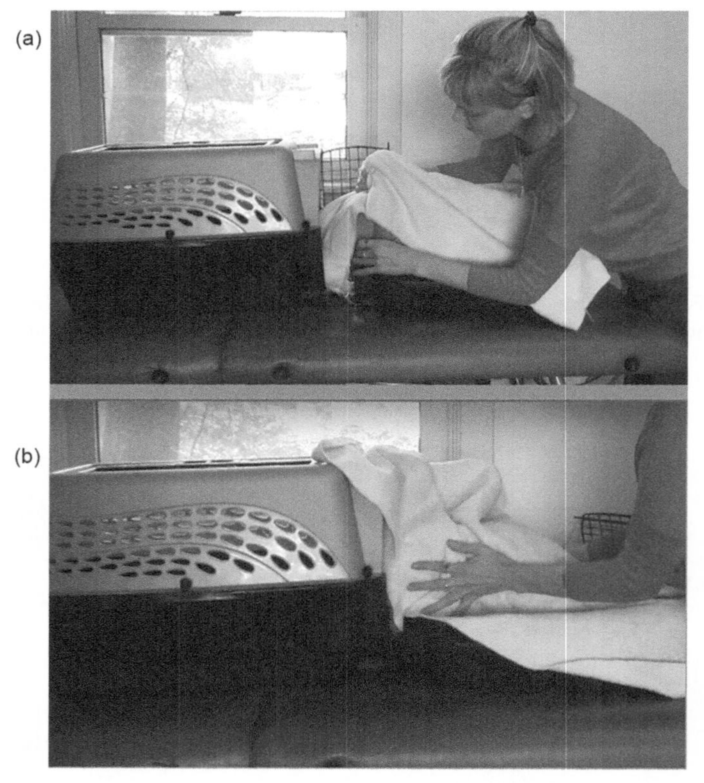

Figure 5.14 (a and b) If the cat won't go into the carrier, remove the top of the carrier the cat is in, cover her with a towel, and scootch her into the new carrier. Use the towel to block off any gaps or openings.

EXAMINING A CAT IN A CARRIER

Often, we take cats out of carriers earlier than we need to and many routine handling procedures can be done while the cat remains in the carrier, especially if there are multiple openings.

If a cat is in a hard carrier and is fearful or huddled, remove the top part of the carrier as you slide a towel gently over the cat. Do as much as you can with her under the towel. If you are by yourself, position the opening of the carrier toward a wall or another barrier before removing the door to the carrier. If there is someone to help you, they can hold the carrier door to prevent the cat from escaping when you take the carrier apart.

Cats can be weighed in the carrier (Figure 5.15). Weigh the carrier with the cat inside and then weigh the carrier once the cat is removed. Alternatively, you can weigh an equivalent-sized carrier. If the top can be removed to the carrier, remove it and weigh the cat in the bottom half of the carrier. Then weigh the carrier bottom, once the cat is removed, or if you don't need to remove the cat and just want to look at her or do some brief handling, weigh the top of the carrier and subtract it from the weight of the cat and carrier bottom to get the weight of the cat. Then put the carrier back together with the cat still in it.

To remove a cat from a carrier with a top opening, drape and tuck a towel over the back and sides of the cat. The towel prevents injury and escape, and the cat doesn't feel your hands directly on her which can frighten her further. Lift her, wrapped in the towel, out of the carrier and move her to another soft surface. If she is huddled or fearful, do as much as you can with her under the towel, moving parts of it only as you need to.

Whether the cat is on a table, your lap, or in a carrier, she should face *away* from you as you handle her. Many cats like being touched on the head, behind the ears, and

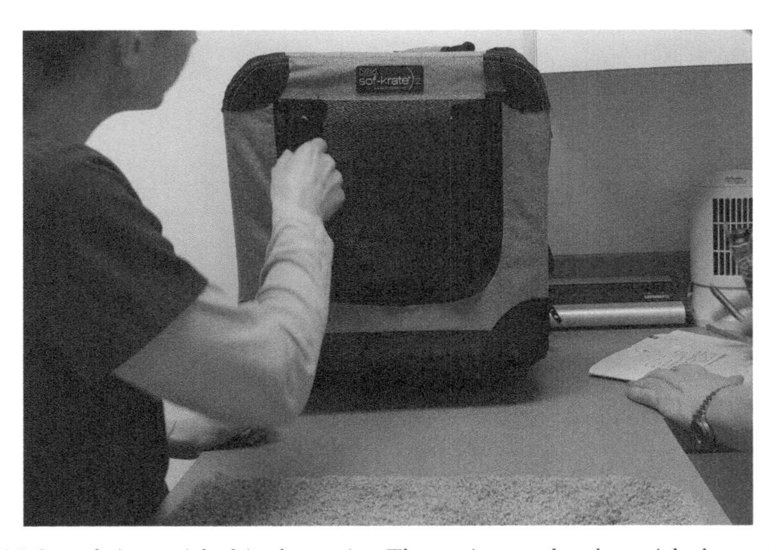

Figure 5.15 A cat being weighed in the carrier. The carrier can then be weighed once the cat is removed.

under the chin, so, if she likes it, gently massage her head and forehead as you look at her. Don't repeatedly tap the cat on the head as a distraction. This isn't pleasant to the cat. Avoid fighting with the cat or repeatedly struggling with her. This will only make cats panic and make them that much harder to handle.

CONTROLLING A CAT'S MOVEMENT

To control a cat's movements, the body (chest and torso) should be fully supported. Cats will often feel more secure when they lean against your arm or body. You can rest your hands on the cat's chest to prevent her from moving forward which can convey to her that you want her to remain still.

Instead of 'scruffing' the cat to immobilize her, use a towel to cover or wrap around her. This way you can protect yourself and the cat as you handle her (Figure 5.16). Another approach is to gently place your hand on the back of her head with your thumb and pinky on each side of the jaw. Your second, third, and fourth fingers can rest on top of her head. If she likes it, you can massage her forehead as you examine and handle her (Figure 5.17).

When holding limbs, move the cat's leg from above the joint while supporting it. Keep the limb within its passive or natural range of motion. When holding hind legs, keep them in their natural position. Don't squeeze the tail with her hindlegs. Let the tail relax. Do not stretch or pull on the cat's legs. Positioning or stretching limbs in ways they do not normally move or rotate is painful. This will make the cat react and she will try to get away to avoid discomfort.

Always start with the easiest and least stressful handling first. More stressful procedures and handling can be done at the end. For instance, removing thick matts,

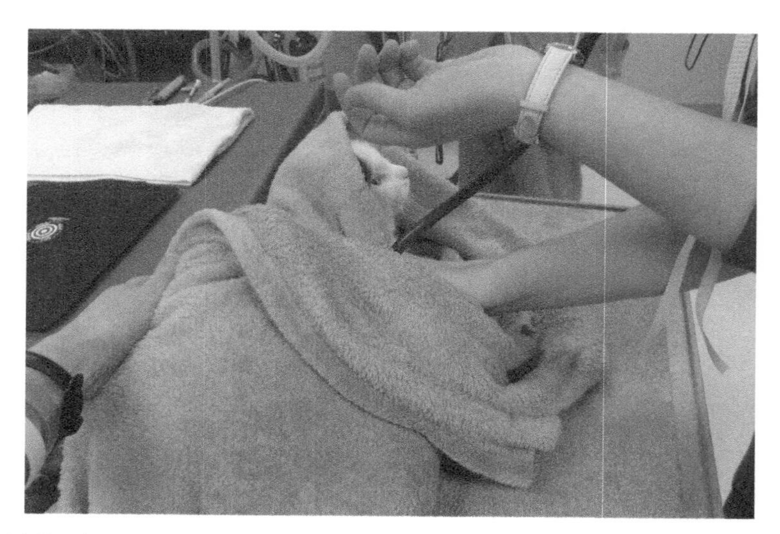

Figure 5.16 Gently cover or wrap a towel around the cat to protect yourself and the cat as you handle her. © Dr. Sheilah Robertson.

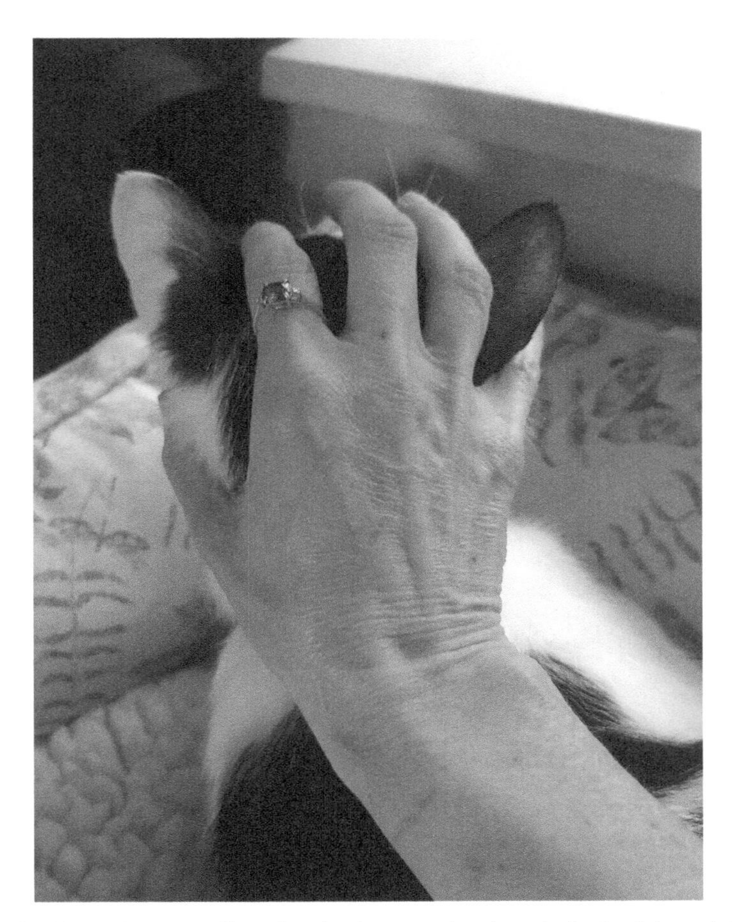

Figure 5.17 An alternative to scruffing. Gently place your hand on the back of the cat's head.

swabbing ears, or examining painful areas should be performed last. Be mindful to give the cat breaks in between procedures. A few minutes rest before continuing to handle the cat can keep her calm so you can start again.

Many cats can be distracted or redirected with food or play while handling them. Kittens can be distracted with toys and food and some adults with highly palatable food. This can take their focus off of handling temporarily and provide a positive distraction for them.

At home, pair any routine procedures with praise, food, catnip, play, or brushing. When looking at your cat's ears, gently massage her neck and chin. Then, give her some treats afterwards (Figure 5.18). After giving oral medication, squirt a little water from a small needleless syringe into her mouth to remove the medicinal taste. Then, give her something tasty or pungent such as tuna juice (water from the can), condensed milk or cream, bits of meat, whipped cream, or wet food.

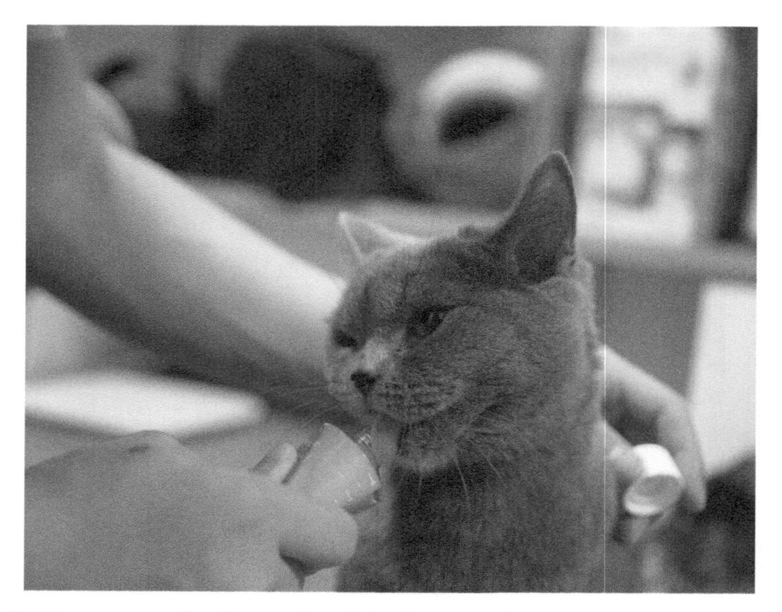

Figure 5.18 Many cats can be distracted with highly palatable food during routine handling and procedures. © River Kao.

PLACEMENT AND HOUSING OF CATS IN CARRIERS AND CAGES

Where cats are placed in carriers and cages can further frighten them or keep them calmer. Whether in cages or carriers, avoid keeping cats near foot traffic or adjacent to slamming doors. Being low to the ground makes cats feel vulnerable and loud noises are scary.

Place carriers on higher surfaces or countertops to get cats off the floor. When housing cats in cages, put fearful and shy cats in the top cages. Avoid using lower cages for cats. Not only do higher cages make cats feel safer and more settled, they prevent escape and making handling as well as moving a cat to and from a carrier or to another cage easier. If lower cages must be used, put less fearful and more social cats in them.

If you have to place a fearful cat in a lower cage, there should be something soft for him to lie down on or a soft domed bed or small carrier to hide in. In addition, if the cat is timid, cover the cage with a blanket or towel so it's less threatening for him.

Small changes and environmental modifications can make a big difference. For carriers, place something soft for cats to lie down on. You can line the bottom of the carrier with medical chux pads (bed pads for people) or with unscented housetraining pads for dogs. These are absorbent and soft. A cat will not be comfortable on just newspaper or plastic flooring.

When keeping cats in cages, provide hideaways and separate food, water, and the litter pan from each other. Many cats prefer to hide when frightened. If they cannot

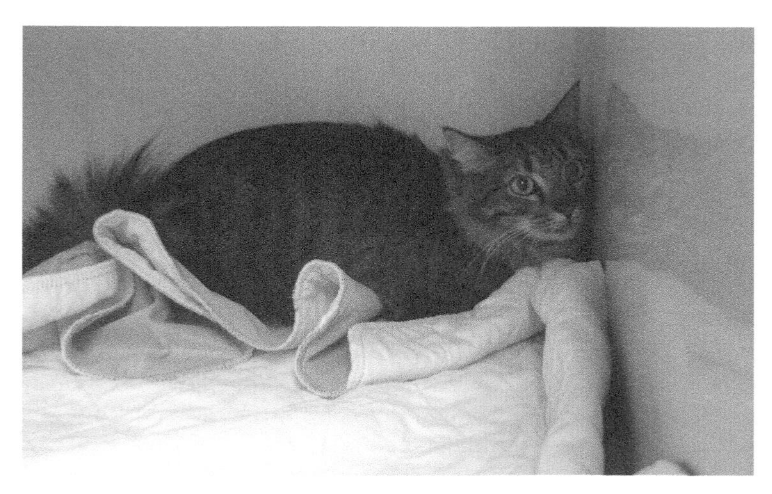

Figure 5.19 This cat is fearful and needs something to hide in. © Dr. Sheilah Robertson.

hide, they can panic or become aggressive. If a cat is huddled in the corner of a cage or lying in his litter box, he needs something else to lie on or a dome to hide in (Figure 5.19). This can simply be placing a paper bag or open box on its side with a soft towel or blanket. If cat beds, boxes, or paper bags aren't available, you can make a cat bed by rolling a large towel into a donut and laying it on top of a soft blanket or another towel.

Food and water should be placed near the cat bed, hideaway, or resting spot, *not* next to the litter pan. Ideally, place the water bowl closest to the sleeping area, food a little further away, and the litter box on the opposite side or corner of the cage.

APPROACHING A CAT IN A CAGE

When approaching a cat in a cage, don't directly face him. Instead, keep your body angled to him. Let the cat approach you first. If he is calm or relaxed, offer your hand for him to sniff and/or give him a treat. Be quiet and speak softly. It's important to avert your gaze if a cat is fearful. If the cat appears frightened or if his pupils are dilated, look at him gently, softly blink, and then look away. Do not lock eyes or stare at him, especially when he's trapped or cornered.

Since most doors and latches on cages are very loud and metal, try to open and close them softly, instead of letting them swing closed behind you or slamming them.

REMOVING A CAT FROM A CAGE

To remove a cat from a cage, introduce yourself to the cat and speak softly, 'Hi Kitty.' Stand or turn sideways to him so you approach him at an angle. When you open the cage door, see if he approaches you first. Spend a few seconds petting or touching him. Position yourself so that you are next to or slightly behind him or turn the cat away from you as you pick him up. Use a towel to drape over the cat, if needed. Your arms, body, and hands should support his body. See 'How to Pick Up and Carry a Cat,' p. 64.

If the cat is in a bed and is fearful or timid, turn the cat bed or position yourself so that you are behind him and he is facing away from you. Remove the cat and the bed together. If the cat gets small or huddles, or tries to hide when you approach him, gently cover him with a soft towel before you remove him with the bed.

GIVING PILLS AND ORAL MEDICATIONS

If there is one thing that many people dread, it is giving pills and oral medications to cats. When we attempt it, the cat either squirms and runs away, the pill gets onto the floor, under the refrigerator, or crumbles in our fingers, or, if we are somewhat successful, the cat froths at her mouth and drools in complete misery. Nobody wants to damage the relationship they have with their cat and poorly executed pilling attempts or giving horribly bitter or unpalatable medication on a regular basis can do just that.

There is a saying 'The medication only works, if the cat takes it.' It's surprising how many veterinary professionals don't taste or sample medication they dispense. Often we are prescribed medication to give our cats that is too sour, bitter, or foul tasting for the cat to swallow, or if it's a topical medication, it burns or stings when we apply it.

If we attempt to put topical medication that stings into inflamed ears or open sores or give the cat horrible tasting liquid to swallow, we are going to fail. Regularly wrestling your cat or forcing medication on her that tastes awful or stings is not a good treatment plan, even if, theoretically, it might work.

If there is a topical medication and if I have a paper cut or scratch, I'll put a dab on the scratch to see if it burns. If there is oral medication, I'll take the pill and dab it on my tongue or taste a drop of liquid to see if it is bitter or sour.

When giving liquid medications and any liquid from a syringe, the cat's head should remain level. When giving pills and tablets, her head can be lifted.

Here are some tips to, hopefully, make it a bit easier for you and your cat, if and when you have to give her medication.

- When a veterinarian explains how to give medication, have them demonstrate it for you. Ask your vet what the medication tastes like or, if it's topical, how it feels when applied. The less offensive or painful it is for your cat, the easier it will be to give to her.
- If you already have the medication and are unsure what it tastes or feels like, sample it. You are not swallowing the medication or taking any dosage that will create problems, so this is harmless and will give you a sense of what your cat will be experiencing.
- Many medications can be 'compounded.' In other words, they can be flavored or modified by a pharmacy to make it less bitter and more palatable. This does not mean the medication will taste good, but it can make it easier for you to give to your cat. For instance, a common steroid Prednisolone is usually dispensed in tablets for the cat to swallow, but it can also be obtained as 'mini-melts,' small vanilla tasting tablets that easily dissolve in liquid. Many oral medications can be given

topically or compounded into a gel or paste that can be applied to the cat's ear. Although this may lessen the medication's efficacy, it's better than not giving any medication at all.

- When you give medication to your cat, initially do it away from her eating area or preferred hang out spots. If the medication tastes terrible and you fail at delivering it, your cat will not have a negative association with her food bowl, mealtime, or her favorite area. If your cat doesn't mind the medication or it's fairly easy to give, you can then give it before mealtimes or wherever is most comfortable.
- Try a pill gun to deliver pills and tablets and practice holding and manipulating it to dispense pills *before* you use it on your cat. Pill guns, if used correctly, can be helpful but can take a little time to adjust to. Once you feel more comfortable and confident with how it works, you will be more proficient when using it on your cat.
- After you insert a pill into the pill gun, add a little bit of butter, fish oil, or meat baby food to the tip. It will make it easier for your cat to swallow the pill and can initially confuse her so she's less resistant.
- Give your cat whipped cream, cold cuts, tuna, sardines, mackerel, or anything else she might love after giving medication. Provide catnip or have a good play session afterwards. If you have to give multiple medications to your cat, give her something pleasurable after each one as well as little breaks in between.

HOLD AND PRACTICE WITH A PILL GUN OR SYRINGE

Practice manipulating a needleless syringe or pill gun before you use it on your cat. One way for more control when working with syringes is to push the plunger with your thumb or palm of your hand, instead of your index finger.

PRE-HANDLING PRACTICE EXERCISES

If you know, at some point, you are going to give pills to your cat, practice a few handling exercises beforehand. You don't have to do them daily, but if you practice in increments on your cat, she can get used to your handling and you can feel more capable. For some cats, you'll be able to go through exercises quickly. For others, it may take more time. You'll progress faster if you stay within your cat's comfy zone (quality over quantity) and give your cat little breaks in between.

Give your cat yummy food after each handling exercise. Follow a similar approach when checking your cat's eyes and ears. Only move to the next exercise when your cat seems comfortable enough for you to do so.

- From next to or slightly behind your cat, with one hand place your fingers on top of her head and your thumb on her lower jaw. Give her treats.
- Hold your cat's head gently in one hand and with your fingers pet or massage her head. With your other hand, touch her lower jaw. Then give her yummy food.
- Hold your cat's head gently. With your other hand, touch her lower jaw and open her mouth slightly. Close her mouth, let go, and give her treats.
- Repeat the above step, but after you open her mouth, gently hold it closed for 1–3 seconds. Give her yummy treats.

- Hold your cat's head and lift it gently with one hand. Open her lower jaw with your other hand. Then let her mouth close, but keep her head lifted as you scratch and massage her throat and under her chin. Give her yummy food.
- Touch and reward with cuddles (and no treats). When you snuggle your cat or she is affectionate or relaxed, massage her head and under her chin. With the same hand, gently touch her upper lip. Then, let go and continue massaging her head and under her chin.
- When she's comfortable with you touching her upper lip, massage her head and face, and with the same hand, lift her upper lip gently with a finger so you can look at her upper canines. Give her yummy treats or go back to massaging her head and face. Progress until you can lift her upper lip above her canines on both sides for a full two seconds ('Good kitty one, good kitty two…'). Let go and give her yummy treats.

GIVING THE MEDICATION

Move slowly, and be steady and gentle. Approach and position yourself slightly behind or next to your cat. Your cat should be in a comfortable position and fully supported on a rug, bed, or sofa, or if in a clinic setting, on a soft towel or mat. Do not lift the cat so she only stands on a few legs or is off-balance. All four of her paws should be on a stable surface.

LIQUID

Give liquid medication from a small needleless 0.3, 0.5, or 1 cc syringe. Keep the cat's head level as you hold her head, and with a few fingers or your thumb, open her lower jaw. Place the liquid medication into the corner or roof of her mouth. *Do not lift the cat's head when giving liquid medications.*

PILLS AND TABLETS

Using a pill gun to deliver pills and tablets can be easier than using your fingers, especially if you practice beforehand.

For pills, lift the cat's head and chin, open her mouth, and with your longest fingers (index or third), place the tablet or pill as far back on her tongue as possible. Close her mouth but keep her head lifted as you rub and massage her under the chin.

All medication should be followed with a little liquid water, tuna juice, or flavored broth via a needleless syringe or in a bowl so the taste of the medication can be removed, and if it is a pill or tablet, it doesn't get stuck in her esophagus.

If you fail at successfully giving your cat the medication, reward her regardless. Try again or in a different way at another time.

INJECTIONS AND SUBCUTANEOUS FLUIDS

At some point, if you have a cat or work with them, you will have to give subcutaneous fluids or an injection. Most injections you give will be subcutaneous, meaning under the skin, as opposed to intramuscular.

There are various ways to hold needles, but for injections, it tends to be easier to control the plunger with your thumb, not your index finger. Often the index finger, which you might be tempted to use first, is too fast and has less control. You can also position the syringe in your hand so your palm controls the plunger.

PRE-PRACTICE EXERCISES ON A STUFFED CAT

Practice on a stuffed animal first. Get a plush, ideally life-sized, cat to practice with beforehand. This will make you more confident.

To hold the skin for injections or subcutaneous fluids, with your fourth and fifth fingers, make a V and lightly grab some skin around your cat's shoulders or on the back of her neck, to the left or right of the spine. Fatty areas will be easier for you and less painful for the cat. With your thumb, second and third fingers, hold the skin to make a tent or divot.

With your other hand, gently, but assertively, insert the needle into the divot you created. It's not necessary to go deep into the skin with the needle, especially if the needle is long. You just need the needle to go deep enough so you can inject the medication. Remember, the medication comes out from the tip of the needle, not from the base.

PRE-PRACTICE EXERCISES ON A CAT

With your pinky and fourth finger, make a V and lightly grab some skin around your cat's shoulders or on the back of her neck to the left or right of her spine. With your thumb, second, and third fingers, hold the skin to make a tent or pocket. Give your cat something yummy or a pile of treats once you let go.

Repeat the above step, make the tent or pocket, and tap or touch the cat's skin with a *capped* syringe or the tip of a *needleless* syringe. Then give your cat something yummy or a pile of treats.

GIVING THE INJECTION OR FLUIDS

Don't use the same needle you practiced with on the stuffed animal with your cat.

Cats dislike abrupt movements, so be steady and quiet, but confident. You will get better with practice.

If you apply alcohol before giving an injection, let it dry before you insert the needle or use witch hazel instead.

INJECTION

Your cat should be in a comfortable position and there should be something soft underneath her.

Feed your cat some wet or moist food, and as she eats or laps it, give her the injection. From next to or behind her, gently grab her skin, make the tent or pocket, and give the injection. Your cat should still be eating after you give the injection.

SUBCUTANEOUS (SUB Q) FLUIDS

Warm up the bag of fluids or saline solution first. Ideally, it should be body temperature. When a drop of fluid is on the underside of your wrist, you shouldn't feel it. Warm it by setting the bag into warm water for 5–10 minutes (do not insert the tube of the bag or the needle into any water) or by resting it on a heating pad, and store the bag in a warm room.

Give subcutaneous fluids at a time your cat is resting or relaxed. The cat should be lying down on something soft and in a comfortable position. From next to or behind her, give her food that she loves. Smear wet food, meat baby food or paste onto a lick mat, give her a bowl of tuna water from the can, a bunch of meat slices, or a small plate of mackerel. As she eats or laps the food, insert the needle. Once you've inserted it correctly, try not to move the needle. If the cat likes touch or to be cuddled, pet or massage her head and face as you give her the fluids.

After providing fluids, play with your cat, cuddle her, give her catnip or something else she enjoys. By pairing giving fluids or injections with something positive for her, the more tolerant she will be and the easier it will be for you.

CLIPPING CLAWS

If you have not clipped a cat's nails before, often small cosmetic human finger nail clippers are easier to work with (Figure 5.20). You would hold them vertically to clip the cat's nails, not horizontally the way we would if we were clipping our own.

Cats' nails grow quickly. When cats' nails get too long and sharp, it can make it difficult for them to play with each other, and if your cat shows any play aggression toward you or lovingly makes biscuits on you, it can hurt. The moment your cat's nails are clipped, you will likely sigh with relief.

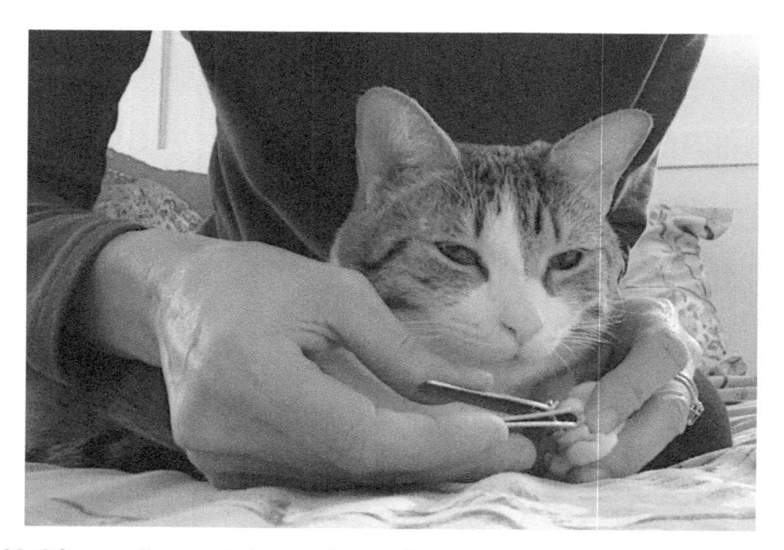

Figure 5.20 Often, small cosmetic human fingernail clippers are easier to work with when clipping a cat's nails.

If your cat's nails get caught on blankets, fabric, furniture, or the scratching post, it's a good indicator that her nails should be clipped. Elderly cats scratch less often, so are prone to ingrown toenails. This means their nails can grow into the paw pad. It's important to regularly check and clip the nails of older cats.

Unless your cat climbs or jumps on your shoulders, the back paws are usually less important to clip. So, if your cat doesn't jump or climb on you, you really don't have to worry about them. However, if your cat is elderly, it will be important to periodically clip his back nails so they don't grow so long they become uncomfortable for him.

Although we like to think there is a magical technique where cats will happily allow you to clip their nails, there isn't. When clipping your cat's nails at home, it is unlikely you will be able to do all of your cat's nails at once. Some cats are easier to handle than others. They may be fine with being placed onto their back in your lap as you clip their nails and touch their paws. However, other cats the moment you even think of touching their toes, look in their direction, or bring out the nail clippers, they're ready to run or swat you.

It might take two or three days or a few sessions to clip all of your cat's front nails, but if you don't overstress him and once he realizes it's not painful, you'll likely be able to do more in one go. Even clipping one or a few nails at a time can be considered a success as long as you are able to stop or transition *before* your cat feels the need to struggle or get away.

PRE-PRACTICE EXERCISES

Use the same approach when acclimating your cat to nail files and grinders.

THE CLIPPERS

If you have a young or new cat, begin by feeding your cat something special as you hold the clippers. Remove or put the clippers away before he finishes eating. If your cat already has a negative association with the clippers, you may need to purchase a different style and/or move where it's been stored to a different location.

Pair special treats and wet food meals with the presence of the clippers. The cat should be aware of the clippers, but not stressed or nervous by their presence. This may mean the nail clippers are placed on the coffee table, floor, or your lap while you give your cat treats or as he is eating.

Next, introduce the cat to the sound of the clippers. Feed your cat as you clip a piece of paper or dried spaghetti. The goal is that your cat shouldn't be fearful of the sound the clippers make in your hand. If your cat doesn't like the sound of the clippers, he certainly won't allow you to trim his nails with it.

TOUCH & HANDLING

When your cat is relaxed, touch his toes gently and give him treats. Frequently hold his toes and paws when you pet him and in between or during cuddle sessions. Don't hold his paw for too long or touch his toes so that he feels the need to pull away. Let go of his toes or stop touching them while he's still comfortable or indifferent. By desensitizing your cat to touching his paws and toes, it should make it easier to clip his nails.

Figure 5.21 Touch your cat's paw and extend the nail when your cat is relaxed and between cuddle sessions. (This cat's claws had already been clipped.)

When you press up on the bottom of each toe, the cat's nail should extend. When your cat is relaxed and comfortable, touch his paw and lift a toe so that his nail appears. Try to do this without having to move or reposition him. For instance, if your cat is on the sofa, instead of trying to turn his paw toward you, sit on the floor so you can view his toes at a better angle (Figure 5.21). Look at each nail closely to see the pink area or nerves of the toe that the nails surround. It's vital never to clip this part.

When touching the cat's toes, try to keep his toes and legs in alignment or within their passive range of motion. This means that the toes, paws, and limbs are moved in the same direction they naturally move or rotate and should be easily moved without resistance. If you move or angle the toe or limb in a direction or way it doesn't naturally go, the cat will resist because it's uncomfortable or painful.

For kittens and cats who like to lie on their backs, sometimes placing them in your lap on their back makes it easier to clip their nails. Practice examining and touching your cat's toes with him on his back on your lap. Touch his toes gradually and gently so you get a feel for how to move and handle his paws without making him uncomfortable. Let go and stop touching his paw before he feels the need to pull it away. Feed him yummy treats while you touch his paw and toes, as well as afterwards.

Once your cat shows no signs of fear or apprehension upon seeing and hearing the sound of the clippers and you can touch or manipulate his toes to see each nail, you are ready to clip his toenails.

CLIPPING THE NAILS

Cats do not like restraint. Clip your cat's nails when he is relaxed and in a comfortable position. By pressing up on the bottom of each toe, the cat's nail should extend. Fortunately, cats nails are clear, so you can see the pink flesh where the nerves of the toe begin. Do not nick or cut this area.

One approach you can take is to clip a few nails while your cat is snoozing and sleepy. Sit or kneel next to or behind him. Gently lift a paw, hold the toe so the claw is clearly visible, and clip the tip only. Do not pull on his legs or paws which will only cause him to pull back in response.

Give your cat yummy treats or food while you clip his nails and immediately afterwards so he can associate nail clipping with something good. If you do not restrain him and are gentle, you may be surprised with how easily he allows you to clip his nails over time. You do not need to clip all his nails at once. Just clip a few a day or every other day.

If you have someone to help you, she can feed your cat wet food, mousse, or whipped cream at the same time and immediately after you clip each nail. Stop clipping your cat's nails before he feels the need to wiggle or pull away. This can be challenging because we often want to take advantage of the cat's patience and keep going. As your cat becomes comfortable, you'll be able to clip more of his nails at one time.

CLEANING EARS

Many cats suffer from ear infections, especially when they have been homeless or out on their own for a long time. Yeast infections are particularly tricky to eradicate and are often overlooked. A cat's ears shouldn't be 'dirty.' The ears should be clean. A little wax is okay, but you should be able see the pinks of the cat's ears and easily see the ear canal. The ears should look healthy and be upright. If a cat scratches or paws at his ears, has 'airplane' ears, or if you see black or brown, thick paste or wax in his ears, they aren't just 'dirty.' He has a yeast infection, bacterial infection, and/or ear mites. Yeast can be very hard to eradicate because only one yeast cell needs to remain for it to self-replicate. Cats can also suffer from hematomas which can make it especially challenging to eradicate an infection because of all the skin folds from their wrinkled ear.

Those who work or volunteer in rescue and at animal shelters routinely have to look at and clean a cat's ears. Upon intake, shelter staff take samples of ear wax or goop to check for mites, yeast, or bacteria. Sometimes, shelter protocol is to clean the ears as well, whether or not the cat has been looked at by a veterinarian. In addition, cats when in unknown environments or handled by strangers are terrified, especially in shelter settings. If the cat's ears look 'dirty,' they are likely inflamed, itchy, or painful, as well.

The 'get it done' mentality is, unfortunately, common when handling cats. Although it might work momentarily, it can make it much harder or even impossible for anyone else to touch or handle the cat. The cat's ears are very sensitive. There is no reason to dig in the ear canal and get huge globs of wax for a sample. Stay around the edges of the ear canal and gently swab as opposed to pushing or digging. Being more aggressive or rough doesn't give you a better sample, nor does it make you more proficient.

When cleaning a cat's ears at home, a little handling more often can be better than a lot all at once. If the cat has painful and itchy ears that need to be looked at and thoroughly cleaned, he should be sedated. This means it should be done by a veterinarian.

APPROACH

Before you clean the cat's ears or apply medication, make sure the cleaner or medication doesn't sting. Cleaners should not be strong smelling and should have a soothing anti-itch, antibacterial, or antifungal component. There is no reason to clean a cat's ears if there is no infection and his ears are healthy.

Position yourself slightly next to or behind the cat. The cat's body should be fully supported on a soft and stable surface. Clean the outer ear with a Q-tip. If you need to remove caked gunk or goo, try not to go into the canal. If the cat shakes her head, the Q-tip can be dislodged from your hand or you may poke at areas you didn't intend to or that may be painful to him. You can use small cotton balls in lieu of Q-tips to gently clean the ears.

Take tiny swabs. Feed your cat treats and give frequent breaks. When applying medication or ear drops, don't pull on the ear. Position the dropper or tube close to the ear canal to dispense the medication. Try to keep the cat's head and ears in alignment instead of unintentionally twisting the cat's head or ear toward you when applying the medication. After applying it, gently massage the ear to move the medication around. Don't do this if his ears seem painful or uncomfortable.

Afterwards, feed the cat treats, play with him, or give him some catnip so he can focus on something else temporarily instead of his ears. At shelters and upon examinations, cleaning a cat's ears, especially if they look 'dirty' or seem uncomfortable should be done last or toward the end of the exam.

WIPING AND CLEANING EYES

Many cats have suffered at some point from an upper respiratory infection or conjunctivitis. This is usually caused by the Herpes virus. Because of this, many cats have dried crust around their eyelids or in the corner of their eyes and can have flare-ups from time to time. Some cats may have runny eyes. If you have any concern about your cat's eyes, she should be seen by a veterinarian.

APPROACH

To clean the eyes or provide warm compresses, use large cotton balls.

Approach the cat from slightly next to or behind her when she is in a relaxed position. Dip a large cotton ball in warm water and then squeeze the water out. Rest the warm, moist cotton ball gently on her eye. Keep it still for 5–30 seconds. Don't press on the eye or rub the cotton ball back and forth as this will be irritating. If you are gentle, many cats remain still. The warmth of the compress and softness of the cotton feels good and is soothing to them, and if their eyes are inflamed or infected, it doesn't hurt (Figure 5.22).

Gently blot the eye with the moist cotton ball to remove any crust or goo. Often, crust on or around the eye will stick to the warmth and moisture of the cotton ball and remove itself. When you are done, repeat the process with a dry cotton ball to dry the fur around her eyes.

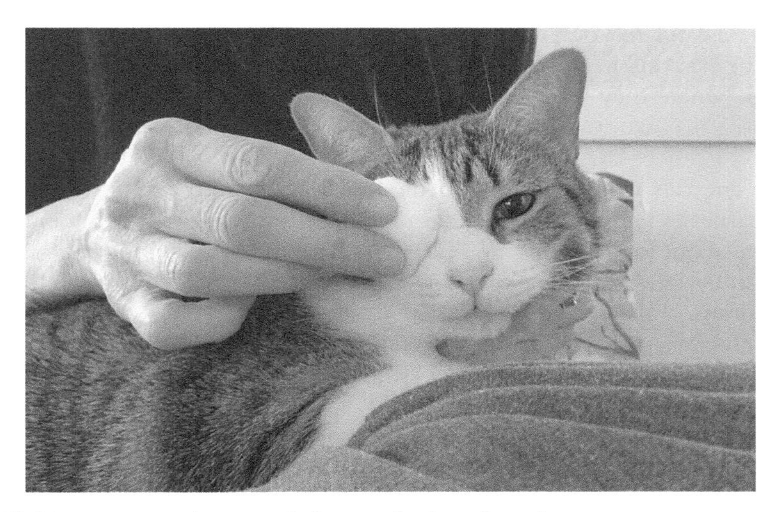

Figure 5.22 Use a warm, moist cotton ball to gently clean the cat's eyes.

To give medication or eye drops, position yourself next to or slightly behind the cat. Lift her chin with one hand and with the same hand open her eyelid. Hold the bottle or tube no more than 1″ above the corner of her eye when you dispense medication. If you drop medication into her eye from too far away, it can hurt when it hits her eye or you can miss and get her fur instead.

After giving medication, gently place a dry cotton ball or soft, clean tissue over the eye for 2–3 seconds. This prevents tears or medication leaking from her eye or dribbling down the cat's face. It also, momentarily, prevents the cat from shaking her head or pawing at her face. Give the cat some yummy food or treats after you have applied the medication. Some cats don't like touch at all. For these cats, immediately let go of them after applying eye drops or ointment.

CATS AND HARNESSES (WALKING)

Some cats like to be outdoors, but it can be dangerous for them to be outside alone. If you live in the country or have a nice yard, you may want your cat to enjoy the outdoors but not take a chance that he will get injured or harmed. If you don't have a yard or quiet space, you may want to get your cat used to a harness as a backup for safety or travel.

Not all cats can be harness trained and not all cats enjoy being outside. Although social media frequently shows videos of cats surfing or accompanying people on hikes, most cats will be far too stressed if put in a similar situation. What you see on social media often isn't reality. It may be choreographed or due to a cat's personality and has nothing to do with the person's handling or training skills. Although you may want your cat to travel the world or, at least, take him to the park, it's likely he wants to watch birds out the window and sleep in the bedroom all day.

Cats startle easily and don't have many ways to defuse stress, so if a cat is sound sensitive, lacks confidence, or is not well-socialized, it's best to take him outside only in a quiet, safe environment.

Walking a cat is not the same as walking a dog. Many cats will sit, nibble on grass (which is healthy and normal), and stay in the same spot for long periods of time. Cats want to go where they want to go. Unless your cat follows you voluntarily, he will likely choose his own way, which might be into the neighbor's yard or in the wrong direction.

Kittens and cats who don't mind being handled and who are more confident can adapt to harnesses fairly quickly. Kittens can adjust to them easier because they are so enthusiastic about food and playing. Cats who have the opportunity to go into a nice yard or garden tend to accept the harness more readily once they associate it with being outdoors.

H-Style or 'wrap around' harnesses tend to be easier to put on cats and for them to acclimate to. They are also harder to escape from. These harnesses have two sets of straps or attachments – one for the neck and one for the chest or torso, so there is no reason to put a cat's legs through any openings or slip anything over his head.

A light scrunchie-styled stretchy leash is good for cats. If one doesn't come with the harness, you can purchase them separately. They are often sold for small animals such as rabbits and ferrets. Cats take to this style of leash better because they are light and they don't feel a sharp tug or sudden jolt when they come to the end of it.

APPROACH

Begin with the harness at a comfortable size. Looser is better initially, so don't make it too tight. Once your cat is comfortable wearing it, you can tighten it if and when you need to.

Expose your cat to the sight of the harness. Let him sniff and look at it. Hold the harness while you feed him high-valued food or yummy treats. You can play with him after he sees the harness too.

When your cat is comfortable upon seeing the harness, progress to clipping and unclipping the attachments and opening and closing any Velcro straps while you hold the harness in your hands. Do not put the harness on your cat yet. Give your cat treats, catnip, or play with him as he hears the snaps and attachments. When he is comfortable with the sounds the harness makes when you clip and unclip attachments, you can begin to put it on him. As you approach your cat from next to or behind him, place down a plate of food for him or a handful of treats. As he eats the food, put one of the straps on him, clip it, and then unclip it. Remove or unclip the harness before he finishes eating.

Progress in increments until you can put the entire harness on him. This should not be done in one session. Your cat should be eating as you touch him and fuss with straps. He should still be eating for at least a few seconds after you've unclipped and removed them as well. Eventually, you will be able to put the harness on him and play with him or give him food afterwards, but make sure he is comfortable with the prior handling first.

If your cat has a high play drive, give him food as you put the harness on him. When he is comfortable with you putting it on, play with him while he wears it using his favorite toy. End the session by removing the harness and giving him a pile of treats.

Once he is comfortable with you putting the harness on, you can attach the leash. Put the harness on your cat and reward him with yummy treats as you clip the leash to the harness. Give him more food or play with him for a few seconds. Then, unclip the leash. The leash should be light. If it is too heavy, it will feel awkward or uncomfortable for him. When your cat is content with the leash being clipped and unclipped to the harness, as well as the feel of it being attached, you can hold the end of the leash as he walks. Hold it lightly and do not prompt or guide your cat with it. He shouldn't feel any tugging or pressure from you.

When you are ready to go outside, pick a quiet time but when your cat is active. Put the harness on him and attach the leash as you give him treats. Open the door to the yard or patio and let him wander outside at his pace.

Once you're ready to take your cat inside, encourage him with your voice and food. If he is playful, try luring him in with a string or pole toy. You may have to carry him inside. Once he enters the home, give him treats or food as you remove the harness. Play with him, give him catnip, or have a good cuddle session. Many cats initially prefer to stay outdoors, so it's important to give your cat something pleasurable after he enters or comes in.

If your cat lies down or won't move when you put the harness on him, don't drag him. Although this might look amusing in videos on social media, it's not productive or healthy for the cat. If you have not yet progressed to going outside and your cat lies down after you put the harness on him, feed him or give him treats as you remove it. Reassess your steps. You might be going too quickly for him or approaching and putting the harness on him in the wrong way. The goodies you are pairing with the harness may not be good enough or you might need a different style harness. When your cat can walk and move with the harness on, then you can attach a leash to it.

VET VISITS & HOW TO BE YOUR CAT'S ADVOCATE AT THE VET

A veterinarian doing a physical exam can be terrifying for a cat, especially one who is timid or has had prior poor handling. Many people don't take their cats to the vet because their cats are so fearful.

Traditionally, the veterinary profession has focused on the physical needs of animals. It has only recently begun to focus on the emotional needs and social well-being of animals. Many veterinary schools don't provide instruction on humane handling or feline behavior at all.

Because the behavioral well-being of cats and humane handling is so often overlooked, it's important to be your cat's advocate. Just because a veterinary professional is knowledgeable about medical treatments, doesn't mean she is gentle when handling cats. One or two bad experiences with a veterinarian or veterinary technician is all it takes for the cat to have a permanent negative association with the vet, the veterinary environment, carriers, or car rides.

Veterinarians and technicians who are good with cats will have a quiet, gentle approach when around cats. Most procedures should be easily performed in the examination room. They will likely do most, if not all, of the exam with you present, including blood draws and giving vaccinations. This may not be the case if special equipment has to be used such as for x-rays, but drawing blood, cystocentesis, injections, nail clipping, and pilling are all standard. There is no reason to take your cat 'in the back.'

Veterinarians who care about cats will recommend and provide ample pain killer if your cat has surgery or if he's in discomfort. They will likely recommend pre-sedation medication for your cat before veterinary visits, especially if they are performing any procedures or if they know your cat is fearful.

In addition, little adjustments can make a big difference for cats. Soft blankets or padding should always be placed underneath them. Gels and lotions can be warmed prior to application. Witch hazel can be used as a topical astringent or disinfectant instead of alcohol because it burns less, or alcohol can be allowed to dry before injections. Letting the alcohol dry beforehand makes the pain from needles sting less. Topical anesthetics can be applied to areas where blood may be drawn, especially for cats who are sensitive to needles.

When bringing your cat to the vet, bring him in a large carrier. When at the clinic, place it on a counter or chair to keep it off the floor. If your cat is frightened, drape a towel or blanket over the carrier. It's also okay to wait with your cat in the car until the veterinarian has a room open for you. At the front desk, request to be called in once a room is available instead of waiting with your cat in the lobby.

During the exam, a soft towel or non-slip mat should be provided for your cat. If one is not provided, request one. This prevents your cat from slipping on a metal or hard table. Cats who are fearful tend to make themselves look small since they want to hide. Not being able to hide can make them panic and become aggressive. If your cat wants to hide and is getting small or cowering, the veterinarian or technician can gently cover his head with a towel. This will keep him calmer, so he remains still. Those who are cat savvy will likely do many procedures with your cat still remaining in the carrier, especially if he is timid or fearful, or they'll remove the top of the carrier and examine him while he stays in the carrier bottom.

If the veterinary visit is routine, for vaccinations or a general checkup, bring some high-valued treats or food your cat likes. Hopefully, your vet will have some on hand. Tuna or bonito flakes, freeze-dried shrimp, cat mousse, and feline greenies are some treats to try. Many cats, especially kittens, can be distracted by licking wet food or with a bit of play while being examined and during injections. If a cat is stressed, he will not be motivated by low valued foods or dried kibble.

If you currently don't have a veterinarian or would like to find a new one, you can begin a search by going to the American Association of Feline Practitioners website and clicking on their Cat Friendly Practice Program. You can also look up Fear-Free Practitioners in the Fear-Free Pets membership directory.

To screen before scheduling a veterinary appointment, ask questions such as which veterinarians or technicians are good or gentle with cats. State that your priority for your cat is gentle handling and for your cat not to be scruffed. It's perfectly okay to request a specific veterinarian or technician, and if you like someone at a clinic, it's important you do so when scheduling. Often it's the veterinary technicians who do much of the hands-on procedures, including drawing your cat's blood. By calling and asking open-ended questions beforehand, you should be able to weed out some clinics by their responses.

6

BEHAVIOR PROBLEMS

INTRODUCTION

Many cats in the US are kept indoors. This is for their safety. But often indoor cats do not have a lot of space or stimulation. They are in smaller homes and apartments and people may be away for long hours at a time.

Many feline behaviors that we consider problematic are often normal, or the personality of the cat, or the age and developmental stage he is in. When we handle normal behavior in incorrect or inappropriate ways, we can create further behavioral problems.

In addition, reality often doesn't live up to our expectations. You may have wanted a couch potato cat, but have a gregarious, playful, outgoing one instead, or you wanted a social companion to keep you company, but now you have a cat attacking your ankles and clawing your arm when you try to relax or walking on your computer keyboard as you try to work. You wanted a cat to enrich your life, but now you can't have friends or visitors over because your cat stalks and hisses at them.

People frequently purchase purebred cats hoping to get a certain personality type. They may think a Persian or Himalayan is good with children or a Bengal will enjoy going for walks on a harness. However, breeders breed for physical confirmation and coat color or what is currently trendy or popular. Cats are not normally bred for their personality or temperament. Often, what people think is a breed-specific characteristic is simply normal cat behavior. There are many variables and environmental factors that influence behavior, but physical appearance generally isn't one of them.

We also tend to be reactive when interacting with animals, thinking that they somehow know right from wrong, can read our minds, or know why they are being punished for something they did. But punishments create conflict and damage the relationship you have with a cat. Added stress weakens a cat's immune system and will make a cat who is already fearful or anxious, even more frightened or reactive. Although punishments and punitive approaches might work temporarily and may be cathartic for us, they will not improve your cat's behavior.

Behavior is modifiable and all animals are unique individuals with their own personalities. You can change behavior but you have to allow cats to be the unique individuals they are.

Many behavioral problems are multifaceted. Most behaviors do not develop overnight and have more than one component to them. Yet, with a little positive insight, patience, and knowledge, most behavioral problems can be remedied or managed.

DOI: 10.1201/9781003351801-6

HOW CATS LEARN

We often focus on what we don't want, which means we focus on what we dislike. But our emotional reactions don't prevent a behavior from happening, nor do they correct a behavior we may not want.

Cats are creatures of habit and learn by doing. The more they repeat a behavior that works for them, regardless of how we feel about it, the more they'll do it. By the time I'm often called to assist with behavioral problems, how the behavior emerged is not important. The cat may be just exhibiting a pattern of behavior. Why is your cat biting your ankles at 10 p.m. at night when you go to brush your teeth? Because it's 10 p.m. and you're going to brush your teeth. The cat isn't putting much more thought into it. Behaviors can become patterns which become routines and habits.

Cats are not self-reflecting (Figure 6.1). They are not analyzing themselves, nor are they looking to you for your acceptance or disapproval in how they behave. How you feel about their behavior is irrelevant. Your feelings or opinions are just that – your feelings. They don't prevent a behavior from happening, nor do they teach the cat an alternate way of behaving.

Cats learn according to the environment or context they are in. I call this their 'bubble.' Any emotions they have will be paired with the context or environment too. Context or situational learning (as I refer to it) happens continuously. We just don't realize it.

The significance of context learning is important when addressing behavioral problems. Where does the behavior occur and in what situations? For instance, your cat may snuggle you in the bed, but she is frightened of you when she's in the corner of the kitchen. Your cat may be affectionate with you when you touch him on his favorite perch or special bed, but if you approach him in any other situation or location, he runs and hides. Aggressive cats can be quite temperamental and can change behavior quickly in different situations or contexts. For example, you might be able to play with

Figure 6.1 Cats are not self-reflecting, analyzing themselves, or looking to you for your acceptance or disapproval with how to behave. © Dilyara Garifullina.

your cat on the floor regardless of his mood, but when you stand up to walk out of the room, he attacks or swats and hisses at you.

We learn contextually, too. It becomes apparent if we're afraid of something. For instance, if you are afraid of big hairy spiders (if you aren't afraid of them, think of snakes), you don't necessarily have to have a reason for being frightened of them, you just are. It's a visceral, negative association you have. If a family member or housemate decides to get a large tarantula or Huntsman spider as a pet and places it in a tank next to the coffee machine in the kitchen, at first you might be apprehensive to enter the kitchen. You might cautiously stare at the tank and startle if the spider moves, or you might avoid the area the tank is in entirely. However, over time, you might get curious and providing the spider doesn't jump out of the tank and chase or bite you, you'll get used to it being there. In other words, you can be in the kitchen, stand next to the spider in the tank, drink a cup of coffee, make breakfast, and maintain conversation. You are fully aware the spider is there, but you are not threatened by it. However, if one day you come home, expecting the spider to be in the tank, and instead you find it on your bed or on the coffee table, you might not know what to do. You're not comfortable with the spider in multiple contexts, nor do you like the spider. You are familiar with the spider in the kitchen and tolerated it in the tank.

Animals who are frightened of people or other animals, and when in new situations, can respond the same way.

Cats learn from each and every interaction they have with us. Cats need to feel safe and care about how we make them feel. Since cats have few ways to resolve conflict, when a cat becomes anxious or fearful, she quickly escalates to freezing, fleeing, or fighting. This is why changing behavior in cats has to be done in a way that makes them feel safe.

Giving cats what they need emotionally and physically, as well as enriching their environment, prevents the majority of behavioral problems.

BEHAVIOR MODIFICATION

Behavior modification is using the principles of how animals learn to change behavior.

Most feline behavioral problems can be prevented or remedied by implementing three principles: prevention and management, desensitization, and counter-conditioning.

PREVENTION AND MANAGEMENT

Prevention and management can be environmental, physical, emotional, or behavioral. Prevention means to prevent unwanted behaviors from occurring and to avoid exposing the cat to anything that might cause her to become upset or to perform the undesired behavior. Since you want to prevent unwanted behaviors, timing is important. It's important to change or interrupt unwanted behavior or emotional states before they manifest or at the very onset. You have to be preemptive or proactive, not reactionary. The better you are at preventing problem behaviors and redirecting them to more desirable ones, the more success you'll have and the quicker you'll see progress.

DESENSITIZATION

If your cat is fearful of something, has never been exposed to it, or is hesitant about it, whether it's a sound, sensation, or another animal, you may want to expose her to it, but need to do so in a way that isn't frightening for her. If a cat becomes anxious or uncomfortable, she'll retreat or panic.

Desensitization means the gradual or incremental exposure to something your cat may not be familiar with or may have had a previous negative association with, but at a level that is comfortable for her. In other words, you can expose her to it, but you have to stay within her 'comfy zone.' Your goal is for the cat to accept what you expose her to so that the next time she experiences it, she is relaxed, content, or indifferent to it. Desensitization is a step up from total avoidance. How we implement desensitization will vary for each individual cat.

COUNTER-CONDITIONING

With counter-conditioning, you want the cat to have a positive experience to something new or that she hasn't been exposed to before, or we want her to *like* something she previously didn't. If a cat is fearful when you look at her, we don't want her to tolerate it, we want her to like when you look at her. If the cat dislikes being carried or stroked, we want her to love when you hold or carry her and to actively seek you out for touch.

We often, unintentionally, make neutral things or what could have been positive for the cat negative. A cat's fearful reaction to the carrier is an example. The carrier was neither good nor bad when your cat first saw it. But since the carrier always appeared before you took her to the vet or you had to wrestle her into it, she developed a negative association with it. Now the moment she hears or sees it, she runs in horror. She is not only afraid of the vet or unhappy with how you handle her, she's afraid of the carrier too.

When implementing counter-conditioning, you want to provide great things for your cat or make her feel happy whenever she sees, hears, or feels what you are exposing her to. How successful you are will be in your timing and delivery. For many behavioral issues, food can be used as something pleasurable, but it can also be play, brushing, baby-talk, or anything else the cat might enjoy.

If we implement behavior modification to change the cat's negative association with the carrier, desensitizing her would be getting her familiar with a new carrier and acclimating her to the sight and sound of it, so she's indifferent to it. She's not afraid when she sees it, nor does she run away. With counter-conditioning, you want her to like the carrier. In other words, if you place it down, she walks into it or she seeks it out as a place to sleep and rest.

SOCIALIZING FERAL & FRIGHTENED CATS AND KITTENS

A number of things can cause fear and timidity in cats. Poor nutrition in the mother during pregnancy or lactation can cause abnormal behavior or developmental problems in kittens. Very early weaning, if also associated with lack of food or poor nutrition, can cause kittens to be fearful. There is a 'friendly' gene in male cats that can be transferred to offspring. So, some behavior is genetic. Some kittens and cats, regardless of environmental stimuli, are more inclined to be shy and timid. Lack of positive exposure to people and handling during the socialization period in kittenhood, as well as prior poor handling or mistreatment, will cause cats to be frightened of people, too.

Figure 6.2 Many cats and kittens hide when in unknown environments or with new people. © Natasza Rusinek.

Many cats, when in new or unknown environments, hide and avoid people, animals, or other cats, regardless of prior exposure to people (Figure 6.2). These cats may be bonded or acclimated to only one or a few individuals, but they behave no differently in new or frightening situations and settings than cats who have had no experience at all.

Shy, unsocialized, or feral cats require many positive experiences with an individual to overcome their lack of exposure or any prior negative experiences they may have had. Highly unsocialized cats can become attached to one person or to a small group of individuals over time.

There is no timeline on fear. *Have patience. When an animal is fearful, he needs time to destress and calm down. Trust is built over time through experience. It is not simply given.* Quiet and skittish animals can come out of their shells and blossom once they feel safe, secure, and confident.

APPROACH

Designate an area or room in your house that is clean, quiet, and safe. Position all food, water bowls, litter pans, cat beds, and platforms so that, when using them, the cat has a nice peripheral view or vantage point of the entries and doorways to the room.

Provide a heat source. Cats love warm, soft things to lie down on. A heated bed or heating pad will provide comfort. Soft sherpa, fluffy fleece throws and blankets can be inviting to cats. When cats are relaxed, they are less fearful.

It's important that cats have the ability to retreat and hide, but they should also be able to see you if you are trying to acclimate them to your presence. Under the bed can be fine, providing it's not cluttered and when you sit or lie on the floor, you can see each other. A frightened cat will never acclimate to your presence if he is hiding and completely out of sight.

Cat domes (enclosed cat beds with entries) or hideaways on cat trees and platforms can provide cats with privacy and safe places to hide while still allowing them to see you. In order for cats to be desensitized to your presence, they have to be exposed to you – albeit at a level they are comfortable with. If they are always hiding so you don't see them, they will never learn you are safe or associate you with good things.

Cats can panic and startle in response to sudden noises and quick movements. This is especially true when they are fearful. We respond the same way when we're afraid. Think of a big, hairy spider suddenly jumping, landing next to you, or scurrying across the carpet. Talk quietly and walk softly around frightened cats. Heavy footsteps will only make the cat retreat.

Introduce frightened cats and kittens to your presence without pushing them to engage or interact with you. Enter the room or area they are in and sit down. Sit on the floor, read a book, have a cup of tea or coffee, work on your laptop, or write and meditate. Look at the cat, verbally acknowledge, and slowly blink at him, then go back to ignoring him. By being present but not forcing any interactions, the cat will feel safer and more comfortable with you.

Talk soothingly to cats. Higher sounds (with the exception of distress calls) are associated with nurturing young and courtship. This is true for most animals. This is why we naturally talk softly or baby-talk to newborns. Lower pitched sounds are commonly associated with aggression or territoriality. Most animals prefer baby-talk and soft, higher pitched voices over normal or loud conversation.

When you sit down near the cat, position yourself at an angle to him. Don't position yourself so that you unintentionally block or trap him by your presence or so that he feels the need to retreat. The cat should be able to leave when he chooses and have a place to retreat to without being cornered.

Whenever you greet or say hello to a timid cat, softly blink at him. Cats softly blink to show gentle motives and friendly intentions. By blinking softly at the cat when he looks at you, you can convey that you are friendly and non-threatening. Don't stare at fearful cats. Staring can be scary. It's okay to look at the cat, but it's important to look away again and break eye contact.

Give the cat highly palatable food every time you enter and exit the room. The cat may be too fearful to eat in your presence, but the food should be associated with you. Wet food, tuna water from the can, bonito flakes, feline greenies, chicken, shrimp, cooked meats, and whipped cream are examples of foods to try.

Add large vertical platforms and window perches to the room. Make these surfaces inviting to the cat by providing intermediate steps so that he can easily access them. Place special food and treats on top of the platforms. Often, fearful cats will climb and sit on these surfaces when you aren't present. They will usually hide when you reenter the room. It's important to get cats off the floor so they feel more comfortable in their surroundings.

Sleep in the room with the cats. Cats are more active in the evening and they feel safer in the dark. Fearful cats will hide during the day, but in the evening, they will come out to explore. By sleeping with them, you can lay on the floor and expose them to your presence, in a way that is non-threatening. However, don't sleep in a noisy sleeping bag. Be aware of any sounds that might frighten the cat and avoid making them.

Some cats who are frightened will allow you to touch them if you move very slowly and show the right body signals. It's important that when you reach toward a cat to do it gradually and avert your gaze. You should approach the cat at angle, positioning yourself next to or slightly behind him. Don't approach the cat directly in front of him or bring your hand to his face while staring at him. When you touch cats, stroke them gently on the back of their neck or along the edge of their back to the left or right of their spine.

Feathers are a great way to get cats acclimated to touch. If the cat likes to play, move a feather along the floor and around his paws. Gradually touch him with it, then move it away again. Once he learns the feather won't hurt him, you will be able to touch him again.

For cats who are too timid to play, reach in with the feather from slightly behind or next to the cat. Don't directly approach him with the feather and touch him on top of the head with it. Since feathers are light, they are a good way to expose cats to touch, but in a non-threatening manner. You can gradually replace the feather with your hand as you touch the cat.

It's helpful to get feral cats used to hand movements before progressing to the feather. Pair all hand motions and body movement with food or treats. It's good to get cats comfortable with hand motions around them, as well. Make little movements with your hand on the floor or rest your hand near the cat as he eats. Before he finishes eating, move your hand away. Once the cat is comfortable with your hand movements, you can introduce touch. Gradually and gently touch the cat. Move slowly, and as you get close to the cat, avert your gaze. If you are using a feather, once the cat is comfortable being touched by it, slide in your hand to replace it. Stop touching the cat or go back to the feather before he startles. This is a process that can take time. The quality of exposure to you for the cat is better than forcing him to interact or engage with you.

TIPS FOR FOSTERING OR TAKING CARE OF KITTENS

If you are fostering or taking care of frightened or feral kittens, keep them together and with their mother for as long as possible. Kittens need the social companionship of other kittens. Mothers of feral kittens are often kittens themselves. Don't mix kittens of different ages. A 7-week-old kitten is too old for a 4- or 5-week-old kitten. The older kitten will play too roughly with the younger one. Kittens should be kept together but within the same developmental stages or age groups.

Little kittens respond to slow blinking, just as adults do. Softly blink at kittens when you talk to them. You will see their eyes soften and they will blink back.

Kittens and young cats love to play. Associate play with your presence. Toys should be quiet and small. It's important to play with frightened kittens in a way that is not intimidating. Don't swing the toy or string toward the kitten, approach him with it, or stare at him while he's playing. Focus on the toy or string, not the kitten, no matter how cute he is. Staring at fearful kittens will only frighten them further. Look at the kitten, but then look away again. End play and any social interactions with food and treats.

Leave small cat toys out such as tiny mice and crinkle balls for kittens to play with when you are not in the room. Rotate toys every day or two to keep them interesting.

Leave dry and/or wet food out at all times for kittens to eat when they need to. Kittens and young cats should be well fed and always have access to food.

INTRODUCING A FEARFUL CAT TO NEW PEOPLE AND GUESTS

Many cats are not well socialized to people. Most people's lives aren't conducive to socializing kittens and young cats. People usually obtain kittens at 7–8 weeks and sometimes younger. Because kitten development goes by so quickly, most kittens aren't exposed to gentle handling by many individuals at the time period they are most receptive to it. Additionally, many people acquire cats as adults after the 'socialization period' in kittenhood has passed.

Cats are selective, and most cats, including kittens, have had positive exposure and experiences with only one or a few individuals. This is why many cats hide when guests come over and are fearful around people they don't know. Prior poor handling or treatment the cat may have experienced only makes her fear more severe.

APPROACH

Don't carry a fearful cat to meet guests or force her to have any interactions with strangers. This will only make her panic. She also won't trust you around visitors.

When guests come to the door, instead of ringing the doorbell, have them knock softly, text, or call you to open it. Sometimes, it's good for a fearful cat to be in another room when strangers arrive. If a person enters a room the cat is in, he should walk softly and speak quietly. Ideally, shoes should be removed, unless the floor is carpeted.

If your cat is in another a room when a stranger arrives, periodically go into the room she's in and give her treats. Give her catnip or anything else she may like. This will establish a positive association for her with the arrival of visitors and hearing their voices. Once people are settled, you can open the door to the room she is in so she can investigate.

Ideally, guests should sit on the floor, especially if they are in the same room with your cat. Minimally, they should sit on a sofa or chair. Standing up and walking around the room frightens timid cats.

It's okay to block access to hiding spots where your cat can completely disappear. If she disappears and remains out of sight, she thinks she is safe because she is hiding and people can't see her. Allow her to retreat, but provide her with hideouts and hideaways so she can see guests or they can see her. People can look at her from a distance and talk to her softly, but then they should look away from her again. Trying to make friends with a timid cat by reaching for her or staring at her will only frighten her.

Slow blinking is a wonderful signal to give to cats, especially upon greetings. Often, if you slowly blink at a cat a few times, even at a distance, you will see her eyes soften and she will blink back. When guests introduce themselves to your cat, they should verbally acknowledge her, say hello softly, and slowly blink at her once or twice. Then, they should quietly look or walk away and ignore her.

Have highly palatable food or treats available for guests to give your cat. When they give her treats, they should look away from her or avert their gaze as they place the treats near her. Then, they should step or walk away from her so she feels comfortable enough to eat the treats.

Kittens, even timid ones, love to play. But, avoid toys that are too large such as large feathers, pompons, or feather dusters. Larger toys can be scary, especially when a person approaches them with it. All toys should be quiet. Remove any bells or attachments that make noise. Guests should sit quietly and move a string or ribbon along the floor.

It's important that people look at the string and not at the kitten. They can glance at her playing, but then they should look away again.

If your cat is curious, let her sniff and investigate people. Guests should relax and allow her to sniff them, without making any quick movements or trying to pet her. If they want to interact with her, they can extend a finger for her to sniff as they softly speak to her, but then they should look away from her.

Praise your cat and reward her whenever she shows confidence or makes any attempt to approach people or explore rooms they are in.

Play light classical music when guests arrive or when the cat is in the room. Soft, classical music can be soothing for cats.

If your cat hisses, people can give her treats, but then they should look or walk away from her. It's important that people stay calm and remain positive or nonreactive. Hissing means that the cat feels threatened or is fearful, not that she will be aggressive.

JUMPING ON COUNTERS AND FURNITURE

Cats naturally like to climb. Being higher up provides cats with a nice vantage point or view of the rooms they are in. Indoor cats have limited access to space. Preventing access to counters and furniture restricts that space even further. When cats are on the floor, they can feel vulnerable. If there is a lot of foot traffic, whether from people, animals, or other cats, it's harder for cats to distance themselves unless they retreat or hide.

Sometimes, allowing your cat on tabletops, counters, and dressers is far easier than fighting your cat not to get on them. Kitchen islands and countertops tend to be in great spots for cats. They are large and sturdy, provide a great vantage point, and tend to be centrally located (Figure 6.3).

If a cat lives in a large home with multiple rooms and floors, it can be easier to keep her off of counters and certain furniture. However, many homes and apartments only

Figure 6.3 Kitchen islands and counters are sturdy, stable, and in perfect locations. © Dietmar Ludmann.

have one floor or a few rooms. The kitchen island or dining room table is often the best cat space in the home.

There is nothing that can make you ill from allowing your cat to walk or lie down on the counter or table. Cleaning the counter and tabletops with all-purpose disinfectant wipes before food preparation or dinner is enough to kill any bacteria, and wiping counters with a damp cloth or paper towel will remove any hair that might be present.

Jumping on kitchen stoves, of course, is a safety issue, but instead of saying 'No' to your cat or squirting her with water, it's important to provide alternatives for her. You can also purchase child safety locks and attachments for the stove.

APPROACH

To remove a cat from the counter, gently pick her up but do not look at her. Then, place her down on the floor, so that she is facing away from the counter. Ideally, pick her up and place her on an alternate surface such as a windowsill, chair, or cat tree. You may have to reposition her more than once, especially if she's young. But, if you're consistent and calm, she will hesitate before jumping up on the counter or table or will stay off of them entirely, at least when you're present.

It's important to provide your cat with alternatives. Add vertical platforms or designate furniture you would like her to be on. If she wants to be in the same room with you, the platforms or furniture you designate should be in that room too. At the very least, she should be within sight of you. The more platforms you have for her to climb, rest, and perch on, the more options and alternatives you'll have when you reposition her.

To teach your cat not to be on the counter during food preparation or while you are eating, designate a location for your cat to sit or rest on instead of the countertop or table. When you are cooking, if she jumps onto the counter, reposition her to your preferred location. This might be a chair or stool next to the counter. If she keeps jumping back onto the counter, place her on the floor or move her to a different room. You may have to do this multiple times, but if you remain calm and provide her with better alternatives, over time she will stop jumping onto the counters or table, or at least, much less often.

RUNNING OUT THE DOOR

Teach your cat to stay away from the door. Designate a spot away from the door to greet your cat. Don't make it too far away. Whenever you enter the front door, walk or carry him to your greeting area and give him treats, affection, and play with him. When you exit, if your cat runs to the door, pick him up and place him on a higher surface away from it. Place him down gently so he faces away from the door. Give him a handful of special treats and then leave. Be sure he has plenty of treats so that he is still eating after your exit. Over time, he will jump up or stay in these areas when you leave for the special treats you give him and will run to greet you in these locations when you come home. As an additional barrier, you can get free-standing dog gates or install a baby gate in front of the door to create a little foyer. This way you can prevent your cat from running out the door as you exit. It also provides you with time to intervene if he jumps over the gate to dash or run past you.

CHEWING

Chewing inanimate items or soft rubber and plastic is normal for kittens and adolescent cats. Just as toddlers explore everything with their hands and mouth, kittens want to taste and chew on novel objects. Some cats may be drawn to rubber and plastic due to texture or novelty – it's interesting and fun to chew on. Even human children like to chew on rubber and plastic. Many plastics and cords are made with tallow or gelatin and some have suggested that this may be a reason cats are attracted to plastic.

Cats have a fondness for electrical cords since they look similar to snakes and worms which make them entertaining to play with. Chewing and tearing things apart is also fun and mentally stimulating. Cats can have a great time chewing and ripping apart cardboard, for instance.

Often cats don't eat what they tear apart, they just like to chew and destroy it. Many cats enjoy biting and chewing on yoga blocks, rubber mats, dolls, dog toys, branches and twigs, paper, rubber bands, shoelaces, and strings on sweatshirts.

Some cats like eating plastic bags. Of course, this is dangerous. Management is the best prevention and giving alternatives for your cat to chew on such as paper, paper bags, wrapping paper, and cardboard boxes. The alternatives may not be as appealing as a plastic bag for your cat, but they're safer.

Cats will chew out of boredom and frustration, and as a displacement behavior such as when they are conflicted or not able to access something they want. For instance, a cat may chew on cords and inanimate objects after another cat takes over a play session or hogs a toy.

Excessive chewing is aggravated by strict diets and meal restrictions, boredom, under-stimulation, and lack of enrichment.

Most cats don't readily ingest what they chew. Some cats do. This is called **Pica** which means the ingestion of non-food items. It is commonly seen in cats who are on restrictive diets or meal restriction and underfed. Anxiety, lack of mental and physical stimulation, age, and genetic predispositions can be factors, too. Certain diseases or medical issues such as anemia and intestinal parasites can cause pica, as well.

What your cat eats, how often he eats it, what he does beforehand, and whether he eats these items in your presence will help you determine the cause so you can begin to change the behavior.

APPROACH

Take your cat to the veterinarian if you suspect your cat's ingestion of non-food based items is abnormal or out of control.

Many items are fine for cats to chew on or rip apart, especially if they are not ingested. Young cats tend to grow out of the desire to chew on things so any excessive chewing behavior will likely be temporary.

The more you prevent your cat from chewing, the more likely he will want to chew. Most cats don't eat or ingest the items they chew on. So, if it's okay for you, then let them chew it. Shoes, toilet paper, wrapping paper, paper towels, card board, edible or cat safe plants, and greeting cards are fine for your cat to destroy. If there are strings such as shoelaces or yarn, monitor your cat so he doesn't eat the string and remove string toys or toys with long plastic or metallic strands when you are not there to supervise him. Chewing on the edges of strings to your sweatshirt or on shoelaces is usually harmless.

Items that are dangerous, such as electrical cords, should be covered with electric cord covers or fastened to the wall. They can be coated with a strong smelling oil such as orange oil or covered in a strongly scented mint wax or chapstick as a deterrent. Unplugging cords during the day and when you are not able to supervise is also a good precautionary measure.

Underfeeding, strict diets, and meal restriction is a primary reason for pica. So, if your cat is excessively chewing or eating non-food based items, incorporate more variety into his diet and free-feed your cat.

Enrich your cat's environment. Add vertical surfaces for your cat to climb on and have interactive play sessions with him. The more you engage your cat and the more interesting his environment, the less likely he will chew on items out of boredom or frustration.

If your cat is selecting dirt, grass, pebbles, or sand to eat, he might need more roughage or cellulose in his diet and a change of food. Purchase or grow edible and cat safe plants (be sure they aren't chemically treated). Cats are especially fond of meadow grasses, oat grass, rye grass, and crab grass. Some cats like eating lettuces, arugula, melon, and corn on the cob. If your cat wants to eat greens, meats, or bread, let him indulge. Often, after a little nibbling or if they eat something too frequently, they get bored.

Be aware that cats shouldn't have onions, garlic, scallions, or leeks since foods in the Allium family can cause hemolytic anemia. Chocolate, grapes, and raisins are also unsafe for cats. Since cats don't have a sweet tooth and aren't particularly attracted to chocolate, it is unlikely your cat will want to eat these foods unless he is underfed, hungry, or on a diet.

FURNITURE SCRATCHING

Scratching is a normal, healthy, and social behavior for cats. A cat will scratch when anticipating something pleasant such as when you wake up in the morning, before breakfast or play, when you enter a room, and upon greetings. Because scratching is a socially rewarding and innately ingrained behavior, reprimanding your cat or squirting her with water, shaking a can with pennies, or yelling at her won't work. She will still feel the need to scratch. If you scold or yell at your cat, or squirt her with a water bottle, she will simply fear you or the water bottle and run away.

Cats have preferences as to what surfaces they scratch on. Cats prefer textured surfaces such as tapestry, heavy textured cotton, Berber, and wicker. In addition, the visual scratch marks a cat makes after scratching encourages her to scratch there again. The more textured, tattered, or shredded your scratching post or furniture, the more inviting it is to your cat! Many scratching posts are made of plush carpet or have glossy faux wood bases. Although these look nice in modern apartments or houses, they are not very appealing for cats.

Cats prefer sturdy, solid surfaces to scratch on. Many scratching posts are too tippy, flimsy, narrow, or small for cats. Often when cats stand on their hind legs, they are taller than the scratching posts (Figure 6.4). If the scratching post rocks, sways, or tips over when your cat scratches on it, she will find a better alternative. Sofas are solid, sturdy, and the perfect height.

Scratching posts and pads should be placed in desirable locations. Cats prefer to position themselves so they can see all entries and exits of the room. They like to have a nice

Figure 6.4 This kitten is already too big for this scratching post. © Willian Justen de-Vasconcellos.

view of the doorways. We tend to place scratching posts and pads in corners, against walls, or tucked behind doors and pieces of furniture. This makes them less appealing to your cat. If you have a great scratching post in an undesirable location, your cat will not use it.

Some scratching posts and pads are much better than others. A good scratching post will be sturdy, made of wood, large, and textured. Vertical scratching posts should be at least 3–4 ft in height or as tall, if not taller, than your cat when she stands on her hindlegs (Figure 6.5). Cardboard and floor scratchers should be as wide as possible. A floor scratcher should be the length of your cat and wide enough for her to fully stand on. Narrow scratching pads are less appealing to cats as are flimsy scratchers that hang on doorknobs (Figure 6.6).

For floor scratchers, heavy Berber, coconut, or sisal mats can be inviting to cats. If you have the means, you can make a vertical scratching post by bolting 2 × 4's together or attaching a piece of wood to a heavy base. You can then wrap the wood with textured carpet. Sometimes, people get creative and attach carpet to their walls. If you want the wall scratcher to be visually appealing, you can place a large piece of carpet in a frame. This way it's fun for your cat and attractive in your home.

Figure 6.5 A good scratching post should be large, sturdy, stable, and textured.

Figure 6.6 A horizontal or cardboard floor scratcher should be wide enough for the cat to fully stand or lie down on. Allie © Paula Lichter.

To encourage your cat to scratch on the pad or post, play with her there by moving a string along it. You can also scratch the post yourself in front of your cat. Frequently, cats will join in. Add some fresh or dried catnip to the post, as well. When your cat scratches the post, play with her and praise her.

If your cat scratches on a surface or area you don't want her to, be calm and unemotional. Gently redirect her away from that location to the area you want her to scratch or divert her attention elsewhere.

To deter your cat from scratching on a previously scratched chair or sofa, trim or cut all tassels and frayed edges she may have created. If the surface is wood such as a table leg or banister, use sandpaper and some wood stain to smooth the area. If possible, reposition the furniture so that the vantage point is less appealing for your cat. Sometimes moving a piece of furniture or item you don't want scratched to another side of the room resolves the problem.

You can try blocking access to areas you don't want your cat to scratch by placing a large potted plant or vase in that location or by putting another piece of furniture there, such as a coffee or end table. This may also be a nice location to place a vertical scratching post or cardboard floor scratcher.

A good way to prevent your cat from scratching furniture is to buy furniture or slipcovers made with microsuede or smooth fabric. Smooth, soft, and plush fabrics are less appealing to cats which make these types of sofas and slipcovers ideal.

As a final caveat, don't declaw a cat. Declawing is the amputation of the distal phalanges or toe bones and is a painful, invasive procedure. Complications with surgeries are not uncommon. When this happens, it can cause lifelong difficulties for the cat. Declawing can create behavioral problems such as general timidity, hiding and retreating, difficulty climbing up and down furniture or cat trees, failure to use the litter box, and biting – often associated with play aggression. In addition, declawing also inhibits play and paw/digit dexterity and mobility. A cat who is declawed cannot grab onto toys or objects easily and declawed cats can get frustrated playing.

ATTENTION SEEKING

Many indoor cats have little stimulation, especially if they are alone during the day or the only cat or animal in the home. This is made worse when they are prevented from sleeping with people at night. Cats are frequently discouraged for climbing furniture and getting on counters which can add to their boredom and frustration.

Cats usually sleep in the absence of stimulation. Even younger cats rarely play with toys by themselves for any rewardable length of time. During the day, when people go to work or school, cats nap. When people come home in the afternoon or early evening, cats wake up to interact. Of course, this is when people want to relax, zone out and watch television, or get ready for bed. In the evening, cats are most active, which exacerbates the problem. By the end of the day, cats are restless, under-stimulated, have pent-up energy, and if not free-fed, often hungry.

When cats are bored, anxious, lonely, under-stimulated, or hungry, they can pace, follow you, cry incessantly, knock thing off of counters, and play aggressively in order to get your attention. Your cat is not purposely trying to be annoying. He's trying to interact and engage with you.

If you get frustrated with your cat, yell at him, squirt him with water, or banish him into another room, his attention-seeking behaviors will worsen, not improve. This can then lead to excessive vocalization and crying during the day and at night. The more you ignore your cat, the worse he will feel. Overtime, he may become withdrawn and less affectionate.

APPROACH

Be proactive. Instead of waiting for your cat to solicit you for attention, solicit him. When he is *not* seeking your attention, pet him, engage him in play, carry him, or pick him up and give him affection. By giving him attention when he doesn't ask for it, you will be initiating interactions more, so he doesn't have to.

If your cat seeks your attention, do not shove him away. Instead, pick him up, put him on your lap, pet him, carry him, or play with him.

Build or add vertical surfaces for your cat to climb on. Cats like to climb and being higher up gives them a better view. The more stimulating the environment is for your cat, the better he'll feel, so the more relaxed he'll be.

Enrich the environment. Provide your cat with edible, cat safe plants to chew on. Place cardboard boxes, paper bags, and tunnels on the floor for him to explore. Leave bunny-kicking toys, as well as small toys, on the floor and rotate them to keep them interesting. Scatter treats or catnip randomly throughout the home.

Play regularly with your cat. Cats like to stalk and hunt prey. If the feather swings over your cat's head or the string chases him, he will lose interest or walk away from it. Play in a way that mimics hunting for your cat. (See 'How to Play with a Cat,' p. 49.) The more mental stimulation he has, the more entertained he will be. After you play, give him some food. This is an easy and simple way to transition him to another activity.

Get a heating pad or bed for your cat. Cats love heat. By adding a few heated beds or pads to your home, your cat will likely relax and nap.

Cats behaviorally do better free-fed (having dry food accessible to nibble at will) since they eat many small meals, frequently, throughout the day and evening. If your cat is on a diet or there is food restriction, he will most likely cry for attention and follow you. If you are conflicted about rationed portions vs. free-feeding, see 'Food for Thought,' p. 156.

If your cat is young and energetic, think of getting another cat of similar age and development.

WAKING YOU UP AT NIGHT

It is typical for cats to be up at dawn and dusk and very active in the wee hours of the morning. Young cats are especially energetic. Generally, cats will sleep when people aren't home. During the day, they can spend an average of 2.5 hours resting and 7.5 hours sleeping. At night, your cat may knock things off of cabinets and tables, eat your hair to get your attention, walk over you or paw at your face, pace and yowl, or scratch under your mattress. Some cats will chirp and vocalize loudly when they hunt, especially when carrying a toy or fake rodent.

Your cat may be hyperactive or excessively vocal at night because he's young, the only cat or animal in the home, bored, hungry, or alone all day without adequate stimulation. In addition, cats are creatures of habit, as are we. Your cat may be waking you up at night because it's simply his routine. The more he performs behaviors at certain times, the more they become a habit and pattern of behavior.

To add to your sleep deprivation, cats don't like closed doors. Cats like the option of being able to enter or exit at their discretion. Closing your cat out of the bedroom may cause him to bang at the door or scratch at the carpet underneath it, which can make it harder to sleep.

APPROACH

Play in the evening with your cat multiple times before bed. After play, feed him a big meal.

Leave food out for your cat to nibble at night. He will be calmer, less restless, and less inclined to start crying for food at 4:30 in the morning.

Kitten or cat proof your room so he makes less noise when he's in it. Designate the top of your furniture and dresser as cat space. Put make up, toiletries, and other supplies in small baskets or tubs. Place ceramic trinkets or decorations on shelves that your cat can't access or place them in a small cabinet.

Add vertical platforms for your cat to climb on and scattered tiny mice, balls, and crinkle toys on the floor to give him something to do at night.

Purchase additional cat scratchers and place them in good locations.

Get a heating pad or heated bed. Cats love heat and it may calm him down.

Place treats and catnip randomly throughout the house at night or put down cat safe, edible plants for your cat to chew on.

In the evening, leave cardboard boxes and paper bags on the floor so he can explore and chew on the boxes or paper.

If your cat wakes you up at night and you want him to settle, bring him into your bed. Pet him while you continue to rest or try to fall back to sleep. Since he is energetic, he will likely leave. But, if you continue to bring him to your bed and snuggle him when he makes too much noise in the room, he will begin to stay on the bed with you or, if he doesn't want to be there, he will eventually leave and play somewhere else.

After bringing your cat to bed with you, if he continues to pester you and won't calm down, calmly remove him from the bedroom. If he claws at your door, install a baby gate or free-standing dog gate as an additional barrier.

Sleep with cotton balls in your ears, ear plugs, fans, or a white noise machine. Take melatonin or a sleep aid.

Over time and as cats age, especially if you allow them to sleep in the bedroom with you at night, they will sleep for longer periods and adapt to your schedule.

PLAY AGGRESSION

Biting and play aggression in young cats and kittens is normal. Young cats naturally want to play and are curious. Kittens will keep you up at night, climb the drapes, and decide you make a great tree limb.

Young cats can be pretty bold in their predatory behavior and mock attacks on people's hands, feet, arms, and other body parts. Play aggression is play or predatory behavior from the cat's perspective. He needs to release pent-up energy and has a desire to hunt.

Kittens learn how to play appropriately by playing with their mother and siblings. Since the socialization period for kittens is 5 weeks to 4 months of age, we tend to get kittens when their play skills aren't perfected yet or just beginning to peak at 7 or 8 weeks.

Classic examples of play aggression in cats are biting your hands and ankles, surprise attacks when you walk down the stairs or around corners, stalking, leaping up at your arms or legs and then running or dashing away, and grabbing onto arms, hands, and legs while kicking you with their back feet. You may see their tails flick or wag frequently (Figure 6.7).

Rough play is common in young male cats, especially if they are singletons (the only cat or animal in the home). Kittens who are weaned too early and did not have access to social play with littermates tend to play more aggressively. Play aggression is worsened when a cat is fed small, portioned meals instead of free-fed, left alone for most of the day, and not allowed to sleep in the bedroom with people at night.

Play aggression is usually directed toward people, but it can be directed toward other cats or animals in the home. This often happens when a young cat is paired with an older cat, two cats have different play styles, or when a young cat is paired with a small dog.

The way we play with cats can contribute to play biting and play aggression. Play should mimic hunting behavior. When cats hunt, they go after small animals such as moths, mice, voles, and sparrows. These animals are tiny and camouflaged. Prey animals quiver and make repetitive small movements or they scurry quickly and hide. The stimulating part of hunting for the cat is stalking the prey. A cat may wait 20 minutes, staying motionless, for just the right moment to pounce. If the prey animal becomes aware of the cat, she will freeze and then dash in a hole, under a rock, or quickly fly away.

Figure 6.7 An example of play aggression in cats is to grab onto your hand or arm and kick it with their hindlegs. © Jon Osumi.

When we play with cats, we tend to repeatedly swing and dangle toys toward them and over their head. Prey never follows or taunts the cat, nor does it try to get the cat's attention. In addition, wands on many pole toys are too short. When the pole is short, cats can get distracted and attack your arm or hand instead of the toy. If we play in a way that bores or frustrates cats, they will lose interest quickly.

Do not punish or scold your cat for playing roughly with you. This includes squirting your cat with water, yelling at him, blowing on him, tapping him on the nose, scruffing or pinning him, stamping your feet, or clapping loudly. These strategies will make play aggression worse. If you punish your cat and are rough with him, he may also lash out or bite you when you attempt to pet him. Punishments for play aggression frequently lead to petting aggression.

APPROACH

If you have a young cat and he is the only cat or animal in the home, consider getting a second cat of similar sex, age, and development.

Do not wrestle your cat with your bare hands. It will be challenging for him to understand how to play appropriately with you. Moving hands and feet under blankets is fine since your cat will stalk and pounce on the hidden movement, not your hands directly.

If your cat looks like he wants to attack you, immediately start playing with him to prevent it.

Avoid tickling your cat on the belly. This may encourage him to bite you. The belly and chest are sensitive areas. Although some cats enjoy having their bellies rubbed, especially tubby cats, others will respond by grabbing onto your hand with their front claws and kicking it with their back legs. If this happens, go limp. If you pull your hand away, your cat's instinct will be to latch on harder. Instead, distract your cat with a toy or movement (move a pen under a piece of paper or a string along the pillow). When he gets distracted by the toy or movement, passively remove your hand. If a claw gets caught in your skin, remain still. Avoid making eye contact with your cat as you carefully remove his claw from your hand. Then, move your hand away. If your cat still wants to play, give him a bunny-kicking toy (large stuffed toys filled with catnip) so he can bite and kick it instead.

When you are sitting down and your cat bites your feet, tuck them underneath you or cover them with a blanket. Passively remove yourself and look away from your cat and then redirect him to a string or pole toy instead.

If your cat attacks your ankles as you walk, freeze. Stop moving or walk very slowly. This eliminates the thrill of the attack. It does not frighten your cat and makes attacking you less exciting (Figure 6.8).

Prevent attacks on your legs and ankles by luring your cat to you with food and placing him onto higher surfaces or position him so that he faces away from you as you walk past him.

Strategically place wand, pole, and string toys so you can easily redirect him when you walk up and down stairs and hallways. If you want to carry your cat to prevent him from attacking your legs, avert your gaze as you pick him up. Hold and carry him like you would hold a football. Then, place him on a higher surface off of the floor.

Figure 6.8 If your cat attacks your legs or ankles as you walk, stop moving. Batman © Caitlin Littlefield.

Free-feed your cat (leave food out for him to nibble at will). If you ration your cat's meals or feed him small portions at select times, his play aggression will worsen. Restricting food in young cats and eliminating the ability to free-feed increases the likelihood and intensity of aggressive behavior.

If your cat looks at you and is in the mood to attack, avert your gaze. Locking eyes with your cat or staring at him will elicit an attack. By casually looking away from him, many attacks can be prevented.

We can lessen the intensity and frequency of play aggression by playing in ways that interest cats. (See 'How to Play with a Cat,' p. 49.) Rethink play as a game of 'Hide the Critter' or 'Hide and Seek.' Make a small obstacle course by placing pillows, boxes, baskets, or other items on floor. Drag a string or toy around, under, and behind the objects. After your cat pounces on the toy or string, keep it still for a moment. Then, when your cat lets go or paws at it, move it away from him and hide it again. After he catches the toy, end play by giving him treats.

Always leave a variety of small toys on the floor and rotate them to keep them interesting. Leave large bunny-kicking toys along your cat's walking path and in his social areas such as in front of the coffee table in the living room or on the rug in your bedroom.

Enrich your cat's environment, as best as you can, so he gets more mental and physical stimulation. Add window perches, scratching posts, cat towers and other vertical platforms for him to climb on, fountains, soft sleeping areas, and tunnels, paper bags or boxes for him to chew and crawl in.

PETTING AGGRESSION & DISLIKE OF HANDLING

Some cats will bite or swat when petted. This can occur when cats are not handled gently or frequently in kittenhood. It can also occur when people use punishments in response to play behavior such as biting, grabbing onto and bunny-kicking your hand, or jumping at and attacking your legs. If a cat is poorly handled, she will dislike any attempt at touch or being lifted, held, or carried.

Cats may also bite, swat, growl, or retreat when you touch them if they are in pain or physically uncomfortable. It's important to have your cat checked by a veterinarian, especially if there is a sudden change in behavior.

In general, cats like to be touched and stroked on the top of their head and forehead (in between their ears), around the occiput (the little bony protrusion at the base of the skull), on their cheeks, and along and under their chin. They also prefer on the back of their neck, around their shoulder blades, to the left and right of their spine, and on their back but at the base or lower portion of the tail.

Cats tend to dislike being touched on their lumbar area (sides of their belly below the rib cage and above their hips), and long, repeated or rapid stroking on top of their back, or being patted on the head (Figure 6.9). Many cats dislike their paws touched, as well. Cats can feel threatened or intimidated when you reach and dangle your hand over their head while you stare at them. Although this is something we do when we think we may get swatted, it's a behavior that makes the cat swat you.

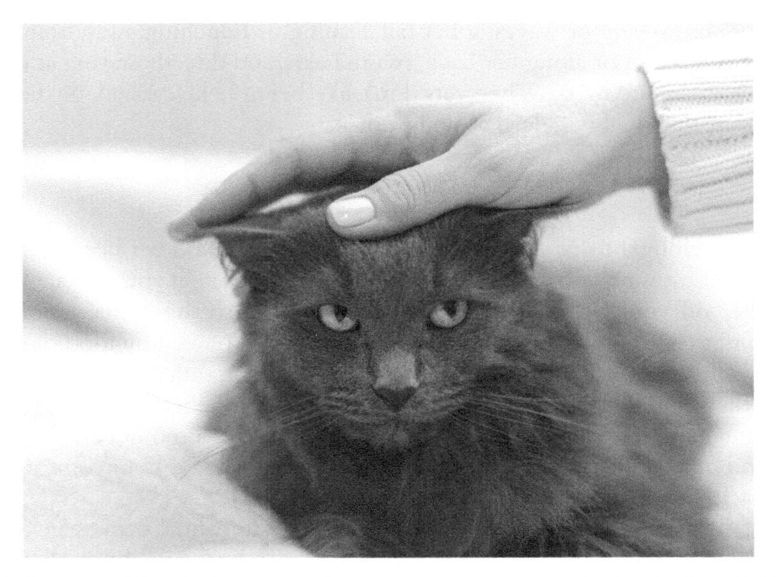

Figure 6.9 Cats can dislike being repeatedly stroked or patted on the head. © Iuliia29photo | Dreamstime.

Figure 6.10 Some cats hiss and swat when you attempt to pet them or after a few too many affectionate cuddles. Vienna © Magaret Hall.

Body language signals your cat will show when she dislikes your touch or how you pet her are flicking or wagging her tail, leaning or hunching away from you, skin rippling or rolling on or along her back, twisted ears, and directly staring at or locking eyes with you (Figure 6.10). When cats don't like being held, carried, or the way you pick them up, they will wiggle, flail, and squirm.

APPROACH

Position yourself next to or slightly behind your cat when you pet her.

When your cat is on your lap and you think she might bite if you attempt to remove her, lure her off your lap with treats or passively move so she is uncomfortable and leaves. Do not try to physically move her unless she is distracted by food or a toy.

Desensitize your cat to your touch and handling. First, touch her only on areas of her body that she likes. Increase her exposure to being touched, petted, lifted, or carried incrementally at her comfort level. Gradually, touch her on other areas of her body for a second or two. Stop touching or petting her before she looks uncomfortable, puts her ears back, stiffens up, or walks away.

For instance, if she leans back, walks, or pulls away after three strokes, next time pet her only once or twice, then let her be. How you touch her, your approach, the movement your fingers make, and where you touch her will make a difference in whether she accepts it or not.

Counter-condition your cat to being touched and handled. Pair anything your cat enjoys with your touch. As you touch your cat – during and immediately afterwards – feed her highly palatable treats. Stop touching her *before* she finishes eating. When she begins to enjoy your touch, touch her in areas of her body you previously couldn't, touch her more frequently, or for longer periods of time. Give her treats immediately after you touch and handle her.

An easy way to acclimate your cat to handling and picking her up is by 'scootching' her to the food bowl. When you put her food down, stand slightly behind her as you place one hand on each side of her waist or lumbar area. Lift or 'scootch' her forward toward the food bowl. After a few repetitions, you will be able to pick her up a little higher and carry her a little further to her food. Use this same approach for anything else she might want. Keep her body parallel to the floor so she is facing away from you and in the direction you are carrying her. Always give her something good after your scootch or pickup. For instance, if she likes a heated cat bed, position yourself behind her and lift her to it. When you place her down, gently rest your hands on her lumbar region and put a pile of treats down for her. Pet her once or twice and walk away *before* she's done eating. Eventually, you will be able to pick her up and carry her further distances.

Let your cat rub herself against your hand and body. When she passes you, rest your hand on her side as she strokes herself against your hand. Glide your hand along her back (to the right and left of her spine) and gently around her tail as she walks by. By letting her fur glide against your hand as she passes, you can acclimate her to touch without reaching for her. Soon you will be able to stroke her when you initiate it.

Don't dangle or hover your hand above the cat's head or hesitate if you want to touch her. Although this is what we tend to do when we think a cat might swat or bite us, this is a behavior that will cause the cat to swat and bite. Instead, position your hand closely at her nose or mouth level, avert your gaze, and slide your fingers or hand up along the bridge of her nose and onto her forehead. Move your fingers on her forehead a few times, then give her food and walk away.

If your cat bites you, make note of where and how you touched her, as well as your manner of approach, what she was doing, and the location she was in. Tweak or alter how you approach her next time so she's more comfortable.

To learn how to pick up and carry a cat who doesn't like being held, see 'How to Pick Up and Carry a Cat,' p. 64.

AGGRESSION TO GUESTS

It's not uncommon for cats to show aggression to guests and strangers, although it's unsettling when they do. Cats can become easily startled or aroused when they're scared. Cats don't know how to diffuse conflict, so it's hard for the cat to understand that the guest or friend isn't a threat. Some cats will stare, hiss, or growl at guests while they sniff, stalk, or follow them. If a stranger stares at the cat or makes a quick movement, the cat may leap at the person's arms or legs or swat at their ankles. When exposing an aggressive cat to visitors, it's important to protect both the guests and your cat.

Don't punish or reprimand your cat for hissing, swatting, or growling. Her behavior will become worse around strangers, not better. It's important to expose her, but to do so gradually and at her comfort level. The quality of exposure to strangers is more important than how long she spends with them. How you introduce her to guests and how they behave around her will influence how she behaves around them.

APPROACH

If you have visitors over frequently, clip your cat's nails. You can begin this while she is napping or sleeping. Start with one or two nails at a time. If her nails are clipped prior to your guests' arrival, it won't hurt when she swats or jumps at them.

Initially, keep your cat in another room when guests arrive. People should call or text you when they arrive at the door or they should knock softly instead of ringing the doorbell or knocking on the door loudly. Once everyone is seated and relaxed, let your cat out to explore.

Have highly palatable food or treats available for guests to give to your cat. In addition, place small pillows, magazines, or 8″×11″ pieces of paper, as well as light blankets or throws in multiple locations so they are accessible to anyone, if or when they need them.

When your cat approaches and looks at the guests, they should avert their gaze. They should not make direct eye contact with her. They can speak softly to her and verbally acknowledge her. They can slowly blink at her, but then they should look away afterwards. They should not attempt to reach toward her or put their hand out for her to sniff, even if she solicits them.

Give your cat a reason to like the guests. Give her highly palatable food when she's in the same room as strangers. If she has a strong play drive, play with her when guests are present. If your cat hisses at anyone, do not scold her. Guests should look away from her and ignore her when she hisses at them. If she stares at the guest, redirect her away from them with food or by playing with her.

If guests want to engage with your cat, they can place treats down for her, but then should avoid making eye contact. If your cat enjoys play, they can move a string around or behind a pillow or another obstacle. They should look at the string and not your cat. Any sustained eye contact can make the cat aggressive.

If your cat intensely fixates or stares at a guest, redirect her away from them. If she doesn't respond or ignores you, the guest can carefully position a magazine, piece of paper, or pillow in front of his face and remain still. Often the cat will stare at the magazine, paper, or pillow, lose interest, and then walk away. You can positively redirect her away, too.

Cats become agitated or startled when people make sudden sounds and quick gestures. If you know there will be activity, bring your cat into another room before guests move about and transition. When guests are back to being relaxed and quiet, let your cat back into the room to visit again.

When your cat seems content with the guests, they can place down treats for her and then stand up, move, or walk away while she eats. Alternatively, feed your cat or play with her and guide her away from them as they move. If she exhibits any threatening behavior or seems uncomfortable, ask the guest to stop moving. Calmly redirect her away from them or bring her into another room.

If your cat attempts to attack any guest or becomes aggressive, that person should remain still and look away. Ideally, have a blanket available so that they can hold it against them if they have to walk past your cat or move away from her. If you can, pick her up and remove her. If you're too nervous to touch her, use a blanket to slowly guide her into another room. When you use a blanket, don't fling it or toss it in her direction. It will just frighten the cat and she will become reactive the next time she sees you with a blanket.

Some cats are so reactive to strangers, they can benefit from an anxiolytic or behavioral medication. Speak to your veterinarian.

INTRODUCING CATS

In outdoor cat colonies, a new cat will stay at the periphery of the group or territory to avoid being attacked until she is accepted. If she's not accepted or there is a lack of resources, she'll leave and go elsewhere.

When integrating cats, do so at a level where each cat is comfortable. Neither cat should feel the need to fight, flee, or withdraw. The distance between cats, duration and frequency of their exposure to each other, their body orientation, your proximity to them, and their energy levels will all be factors when introducing them. Stay within each cat's comfy zone. You will go faster and have more progress that way. The quality of exposure is more important than the quantity.

Cats who are nonaggressive and confident will often look at other cats and walk away, move to the side, sit and groom themselves, or casually look away. Cats who are fearful will cower or hunch in the presence of other cats, with pupils dilated. They will run or try to get away if pursued. Unfortunately, with an emboldened cat, a fearful cat's scurrying to leave the room can entice him to chase. Cats who are more aggressive will assertively stare at other cats without breaking eye contact.

DESIGNATE A ROOM

Designate a room for the new cat. This room should have a litter box, food, and water bowls. Place food and water bowls in slightly separate locations and keep them away from the litter pan. Add scratching posts (both vertical and horizontal). If you don't have any elevated territory in the room such as a bed or sofa, add elevated resting areas and window perches.

The new cat should be put in a spare room, office, or guest room. This is especially important if the resident cats sleep in the bed with you at night. Do not kick resident cats out of preferred locations, especially the master bedroom.

ADD FELINE RESOURCES TO THE HOME

Add extra water bowls, food bowls, litter pans, edible plants, cat trees and condos, window perches, and cat beds throughout the house. Place litter pans in more than one location. Create and add as much vertical territory as you can. One way cats avoid conflict with each other, but can be in the same room together, is by climbing on higher surfaces. If there is nowhere to go but away or underneath things, a cat has no choice but to hide or retreat.

SCENT AND EXPLORATION

Start by mixing and exposing cats to everyone's odor, including yours. The goal is to create a communal scent. There are a variety of ways to do this. Use a cloth or bandana to gather scents from all the cats – both the residents and newcomer. Introduce them to this 'combined' scent by rubbing your legs with the cloth. Rub it on furniture and table legs throughout the home, especially areas your cats may rub on. This familiarizes them to each other without direct exposure. Praise your cats and give them treats or immediately brush or play with them after they sniff any cloths, furniture, or clothing.

If there is a lotion, perfume, or cologne you frequently wear, you can then dab it on your hands, wipe your clothing, and pet each cat.

When all cats are accepting this combined scent or odor, which means they are either ignoring or rubbing against any objects marked with it, then it is time to allow the new cat the opportunity to explore the rest of the house.

When you are ready to let the new cat explore, close the resident cats in rooms where they are already comfortable. Let the new cat explore the rest of the house or apartment. After the new cat has explored and is relaxed, bring the new cat into another room and let the resident cats explore the new cat's room. Let cats explore at their own pace.

If you have multiple cats, sometimes it's good to have each cat explore rooms separately. Pair or end explorations into new areas with food, praise, and play.

When the door to the new cat's room is closed, if resident cats hover outside the room, entice them away from the door with treats, praise, or play or pick them up and reposition them. Try to avoid allowing cats to play footsies or paw at each other vigorously through the door. Often people think the cats are playing, but they may actually be fighting or antagonizing one another. Some light pawing is fine, but interrupt cats before they start fixedly pawing at each other under the door.

Once the cats are using all the resources in the home confidently and not intensely stalking or pawing at the door to the new cat's room, then it is time to visually introduce them.

VISUAL INTRODUCTION

To begin introductions, the cats should be able to see each other without any risk of physical altercations. I recommend using extra-tall baby gates with swing doors to separate the cats (Figure 6.11). These walk-through gates can be placed in multiple rooms so you can expose the cats to each other in a variety of areas. You can also use temporary screen or glass doors. Make use of vertical surfaces, as well.

Watch your body language and the energy you give off. If you are stressed, hovering, frantic, or reactive, the cats will pair it with each other or your presence, not their behavior.

When cats look at each other, say their names and give them treats. Give them highly palatable food when they see each other and for a few seconds after the exposure. If you have a cat who is not food motivated, praise him, brush him, or give him catnip or something else he enjoys. The good things you provide do not have to be the same for each cat.

Figure 6.11 It's good to use extra-tall baby gates or free-standing dog gates, initially, when introducing cats. Ellie & Flower © Audrey Fox.

If you are by yourself and using a baby gate or vertical surfaces, give food to one cat and then give food to the other. Go back and forth, praising them and giving each one something they like.

When feeding cats communally, in other words, at the same time, give each cat enough space so they can position themselves comfortably and eat without worry. If any cat hesitates, they are not yet comfortable.

Defuse any mild threats or conflict, such as staring, swatting, or hissing with positive interruptions or redirection. The goal is to defuse the situation in a positive way and change both cat's emotional states to a more positive one. It's important to interrupt any cat from staring at or stalking another (Figure 6.12). Try to do this in a relaxed and positive way before the other cat feels the need to hiss, swat, or retreat. If positive interruptions or distractions don't work, you can reposition the cat who repeatedly stares. To reposition him, pick him up and place him on a surface so he faces away from the other cat.

Every exposure is a learning opportunity for the cats. Where you left off is where you begin the next time they see each other. Try to end their exposure to each other in a neutral or happy way.

Figure 6.12 It's important to interrupt any staring between cats. Binx & Floyd © Grace Yuen.

When the cats seem comfortable with each other and you are able to supervise, leave the door open to the new cat's room for longer periods of time. Still use gates to separate them. You can double stack gates, if necessary, and attach bells at the top of them so you can intervene if you hear any altercations or if a cat attempts to climb or jump over the gates.

Continue closing doors and letting cats explore all rooms of the home separately so they can familiarize themselves with each other and their shared space. If one cat hides or seems timid, boost his confidence before continuing any further introductions.

When the cats show no signs of fear or aggression toward each other such as staring, stalking, hissing, growling, or crouching, you can have them meet without any barriers between them.

FACE TO FACE INTERACTIONS AND MEETINGS

For face to face meetings or meeting without any barriers, when there are multiple cats, introduce them to each other separately so that introductions are more manageable and not so overwhelming.

Be positive and relaxed. Try not to be frenetic or show nervous energy. It's best to introduce cats to each other in settings or situations they'll be more successful based on their individual preferences and personalities. For instance, if one cat likes food and to be on higher surfaces and the other cat likes to play, recreate that scenario when you introduce them. While one cat is on a cat tree, give him treats as the other cat enters the room. On the floor, play with the cat who entered while you give treats to the one on the tree. If all cats like food, feed them together. Position cats at a distance where they are comfortable and no cat should be cornered or trapped if or when another cat wants to leave.

It is fine for young kittens and hungry cats to eat each other's food, especially if the other cat allows it and doesn't seem fearful. If you see this, give more food to both cats or place them in slightly different locations so there is more distance between them.

Transition cats in a neutral or positive way such as luring one cat out of the room with a toy while the other cat eats or by opening a window for one cat to look out of while you entice the other into a different room with food.

Be positive whenever the cats approach each other, sniff each other, and pass each other. Praise and reward them, but try not to hover.

When you are not there to supervise, separate the cats until you are confident that they can get along with each other or coexist. Once they are comfortable with each other, you can leave them alone unsupervised.

TROUBLESHOOTING

If one cat hisses, it means he feels threatened or is upset. Make a note of when and where he hisses to troubleshoot and verify the cause. Always give treats or good things to the victim cat (the cat who wants to avoid conflict) after any altercations. This changes the victim's emotional state to prevent him from becoming more fearful or anxious.

If there are any altercations such as swatting or batting at each other or quick chasing, interrupt it in a positive, unemotional, or casual way by distracting the aggressor and guiding him away or repositioning him. To reposition him, pick him up and place him on a higher surface so he faces away from the victim. You may have to do this a few times, varying where you place him down each time, but it often works.

If cats are still not getting along or you don't feel comfortable with introductions, go to 'Inter-Cat Aggression and Conflict,' p. 119.

INTER-CAT AGGRESSION AND CONFLICT

If you have male cats who are not neutered, they will fight and spray. Neutering should be the first priority.

Do not punish cats who are in conflict with each other. Shake cans, squirt bottles, yelling, stamping, and loud noises will make aggression worse, not better. The cats will associate the punishments with each other and/or the location they have conflict in, not their behavior.

Cats who are bonded or like each other can share the same spaces or locations at the same time. Cats who are more tolerant of each other may share a larger area, but avoid sharing the same spot as another cat, or they may time-share. This time-sharing may be situational or at set times of the day. Some cats don't like to share at all.

Fights can be between males and females or between the same sex. Aggression tends to increase in the morning and evening, as well as before mealtimes.

One of the most common causes of inter-cat aggression is when one cat in the home is introduced to another cat and they don't get along. Cats generally avoid conflict by creating and maintaining distance from each other. A more confident, assertive, or aggressive cat will chase another cat away or take over the resources. The other cat retreats or leaves to avoid conflict. When cats live in the same home, especially in smaller spaces, they are regularly forced to be in close proximity to each other. If the victim has few places to retreat to, fighting can ensue. Unfortunately, after a fight, neither cat has an opportunity to leave, so fighting continues.

Another reason for inter-cat conflict is when two cats have differing play styles or are in different developmental stages. Often this is when young male cats, who like to wrestle, are paired with older cats and female cats who want to relax or prefer to chase toys, strings and watch birds at the window. Just like play aggression can be directed toward people, it can be directed toward other cats. Although the aggressor may not be purposefully trying to torment or displace the victim, the victim doesn't appreciate being pounced on and chased.

Inter-cat aggression may occur when one cat goes to the veterinarian and upon return, his housemate doesn't recognize him. This can happen with bonded cats, as well. Prior to the visit, the cats were nuzzling and sleeping together, but afterwards, they can't even be near each other or in the same room.

Redirected aggression is yet another reason for aggression between cats. Redirected aggression generally describes when a cat gets agitated enough by a stressor, such as another cat outside or a loud, unexpected noise, that he targets and attacks the nearest victim. This can be a person, dog, or another animal, but it is usually another cat in the home. When this happens, it is traumatic for everyone.

SIGNS OF AGGRESSIVE BEHAVIOR

A *direct stare* is usually exhibited by the aggressor. If you see one cat stare at the other cat with intensity or laser-like focus, such as when spotting a prey animal, you will likely have only a few seconds to intervene.

Passive staring occurs more regularly, but it's often overlooked. This is when one cat stares or looks at the other cat for longer than three seconds without breaking eye contact. I call this the 'creepy' or 'creeper' stare. Cats can use passive staring to displace and intimidate other cats. Because the aggressor looks innocent from our perspective, we tend not to notice it. Some people think the starer might want to make friends with the other cat. However, the cat being stared at looks nervous and uncomfortable, may avoid eye contact, hiss, or growl, and usually leaves the room. Once the victim has left the area, the 'creeper kitty' or passive aggressor will take the spot the victim was in or will follow the victim, situate himself nearby, and then stare at the victim again (Figure 6.13).

Victims of passive aggression try to avoid the staring cat. They may withdraw and spend most of their time in another room. If there are not enough places or rooms to retreat to, they may spend their time in one or two locations, such as in the basement, on top of the refrigerator, underneath the bed, or in a closet.

Staring inevitably leads to stalking, which leads to chasing. This may be subtle and happen over time or this can happen relatively quickly after cats are introduced. Pinning and tackling are two additional behaviors that can either be associated with direct aggression or play.

A nonaggressive cat will generally move to the side, sit, or casually look away from another cat. A cat who is nonaggressive and trying to prevent or avoid conflict may look away from the other cat or slowly blink in that cat's presence.

STEPS & APPROACH

Depending on the extent or severity of the aggression and how often fighting occurs, other approaches and behavior modification suggestions may be recommended, or it might be best to keep the cats separated.

Figure 6.13 Cats can use staring to displace or intimidate other cats. The victim usually avoids eye contact. Winston & Moose © Rachel Lally.

If one cat is so fearful that he continually hides or seems unhappy, it is best to keep him separated from the aggressive cat until he can de-stress and move freely about the home without worry. Once the victim cat is comfortable and relaxed, you can begin reintroductions. These reintroductions are not very different from introducing cats for the first time. (See 'Introducing Cats,' p. 115.) When you are not there to supervise, the cats should be separated in comfortable locations.

The goal of all behavior modification is to prevent and interrupt staring by the aggressor before the victim feels the need to flee, while you establish or create a positive association between them and boost the victim's confidence or give him 'bounce back.' The victim cannot be chased or attacked. You can prevent and interrupt staring and stalking by the aggressor with positive redirection and repositioning.

When integrating cats or reintroducing them, do so at a level where each cat feels safe. Neither cat should feel the need to fight or hide. The distance between cats, duration and frequency of exposure, their body orientation to each other, your proximity to them, and the cats' energy levels are all factors when exposing them to each other.

Cats can have access to all the rooms of the home, just not at the same time. Rotating rooms or areas the cats are in does not need to be for long periods of time. It can be for 15–20 minutes to a few hours scattered throughout the day.

Go around the home and see what resources can be added for the cats to prevent conflict. Make note of where the preferred spots are, the water bowls they drink from, posts and pads they scratch on, and litter pans they use. Add extra water bowls, food bowls, cat trees, shelving, window perches, litter pans, scratching posts, and cat beds to as many rooms as you can. Make sure litter boxes are placed in more than one location, and if your home has multiple floors, on more than one level. Add vertical territory for the cats so they can share space, yet maintain a comfortable distance from each other when they are in the same room together.

Position resources so that cats can see all entries and exits of the room when using them. This is especially important for the victim. The victim should have multiple ways to get away from the aggressor or retreat without being trapped or cornered. This might mean moving a food or water bowl 10 inches away from the wall or positioning the litter pan at a different angle (Figure 6.14).

If there are specific areas where there have been fights or confrontations, look at the positioning and number of resources. Is one cat trapped by the other when they enter or exit the room? Is a cat cornered if he uses the litter box? Do multiple cats have to use the same resource, in the same location, at the same time? Make any changes you think are needed.

Figure 6.14 Altercations can occur when cats have to pass each other or if one cat is blocked, trapped, or cornered with few exit strategies. © Jonnie Turpie.

When the aggressor is absent, invite and encourage the victim to be on higher surfaces. Do not invite the victim to be on a surface that is a preferred location for the aggressive cat. Being on higher surfaces can build a victim cat's confidence, and if the victim feels comfortable enough, he can get away from the aggressor without having to leave the room. When cats are fearful, they tend to hide, scurry, and run away. This can encourage an aggressive cat to chase them.

When reintroducing cats, do not force or coerce the cats to interact with each other. Forcing cats to interact when they aren't yet comfortable will only frighten them or make them more adversarial or fearful of each other.

Passively position the cats in a way that the aggressor does not face or block the victim. The victim should not have to 'run the gauntlet' to get to resources or pass by the aggressor to exit a room.

Initially, use stacked baby gates with swing doors (walk-through gates) to expose the cats to each other in a safe way and so you can intervene, if necessary. When the cats see each other, be happy, calm, and positive. If you are stressed, your cats will be stressed and they will associate that stress with each other.

Feed cats extra palatable food and special treats whenever they see each other.

Give the aggressive cat treats, catnip, praise him, pet, and play with him whenever he ignores, looks away from, or slowly blinks in the presence of the victim. Acknowledge and reward any calm or friendly behavior shown by the aggressor such as keeping a personal distance from the victim, waiting patiently, and turning or walking away from the victim.

Interrupt staring by the aggressor at its onset by redirecting him with toys, play, food, catnip, or by petting him. If the redirection doesn't work, reposition the aggressor. It's important to do this before the victim feels the need to flee or retreat. To reposition the aggressor, unemotionally pick him up and place him on a higher surface so that he faces away from the victim. Initially, depending on his aggression, he can stay in the same room or general area as the victim. Alternate the platforms you place him down on. For instance, if he stares at the victim and positive redirection doesn't work, pick him up and place him on the back of the sofa. If he goes back to staring at the victim, pick him up and place him on the dining room chair, then on a windowsill, then on a cat tree, and so on. You may have to reposition the aggressor multiple times, but the behavior of staring and stalking usually subsides. If the aggressor still tries to stare at or stalk the victim after repositioning, place him in a separate room so he calms down. Then try again at a different time.

It's vital to prevent the aggressive cat from chasing or following the victim whenever the victim decides to leave the room. You can do this by feeding the aggressive cat or redirecting him with a toy at the same time the victim transitions and leaves.

Reward the victim cat for being confident, not retreating, playing, entering areas he formerly avoided, passing the aggressive cat, and slowly blinking in the presence of the aggressor.

Ideally, transition cats by ending interactions in a positive way and when there have been no altercations between them. Exposure is a period of time where the cats learn how to coexist and be around each other. Where you left off in that exposure is where they begin the next time they see each other.

It's important to change the context of any locations or scenarios where fights have occurred. I call these areas 'hot spots.' Conduct play sessions, baby-talk, and/or feed the cats highly palatable food in any areas where there has been confrontation or in any place where the victim has been startled or attacked. You might have to change or move furniture around too. If you cannot change the association environmentally or behaviorally, spray Feliway in the locations or, if there is an outlet, plug in a Feliway diffuser. Feliway may change the negative context of the location to a more neutral one. Do not place Feliway near or next to any litter boxes.

When cats are friendly, which means they can be in the same room together and ignore each other or exhibit normal behaviors such as grooming themselves, drinking, eating, using litter boxes, entering and exiting rooms, and jumping onto and getting down from furniture and vertical surfaces, including windowsills and counters, you can begin leaving them unsupervised for brief periods.

Only leave the cats alone together when the aggressor has shown no signs of aggression to the victim or when the victim is comfortable enough to jump up and down from higher surfaces and walk away from the aggressor without being pursued. If the aggressive cat is unpredictable or too quick for you to interrupt and redirect, you may have to acclimate him to a harness and leash or keep the cats permanently separated for the victim's protection.

Some people can provide two or more separate 'homes' for their cats within their house or apartment. This means their cats are always separated and this may be the preferred or most acceptable outcome for everyone.

IS IT PLAY?

INTEGRATING OLDER, TIMID, AND SEDENTARY CATS WITH YOUNG, ACTIVE ONES

Feline play involves predatory behavior. Cats can become mock-predatory targets for other cats.

It is helpful to understand the play styles of each individual cat in the home. Predatory or aggressive play is common when there are large age gaps or vastly different play styles between cats.

Older, timid, and more sedentary cats can be stalked and chased by younger ones due to a young cat's pent-up energy and intense play drive. Young cats, especially kittens, want to socialize and interact with other cats. They tend to be 'busier' than older or more sedentary cats, need frequent entertainment, and are easily bored. This can lead to one cat chasing the other.

Cats can have different play styles. Young male cats and kittens prefer 'wrestling' with each other, while older cats and females prefer stalking and chasing feathers, strings, and pole toys. This can lead to miscommunication, conflict, and fighting.

When a cat is aggressively *playing* with another cat, you will see one or more of the following: galloping, surprise attacks, getting low to the ground, stalking and/or wiggling the bottom, pouncing, or tackling and pinning the other cat and then dashing or running away (Figure 6.15). At times, the victim may get along with the aggressor. They may even groom each other in calmer situations. Play-aggressive cats can usually be distracted, or if the victim becomes adversarial, they will back off. After a short break, however, they go back to chasing, tackling, or pouncing again.

Figure 6.15 Play aggression can occur when there are large age gaps or differences in play styles between cats. © Chanika Nuwanno.

APPROACH

Expose the cats to each other when the young, active one is busy or engaged in something else such as during mealtimes or when napping. Play with the young, active cat either beforehand or in the other cat's presence so that his attention can be directed elsewhere.

Encourage older, timid, and more sedentary cats to get on higher surfaces so they can distance themselves or get away from the aggressor. Add cat trees, window perches, and shelves, and provide easy access to counters, tables, and dressers. Likewise, by adding vertical surfaces, you have an opportunity to reposition the play-aggressive cat without having to remove him from the room.

When the play-aggressive cat sniffs or tries to interact with the other cats, praise them, give them treats, or do something else they enjoy. Then, redirect the active or younger cat away from the older or more sedentary one so he can focus on something else. Shower the older cat or timid one with food, praise, catnip, or anything else she may like, after any interactions, including being pounced on.

If the play-aggressive cat continues to pursue and pounce on the victim, reposition him. Pick him up and place him on a higher surface so that he faces away from the victim. You may have to do this multiple times, changing where you place him down each time. If he continues to pursue the victim, put him in another room temporarily so he calms down.

Once the cats can sleep in the same room together and the victim can easily and confidently use all resources in the home, you can leave them alone, unsupervised, for short periods of time. When you are confident they are comfortable with each other and can coexist, you can leave them together unsupervised.

Play aggression can last a few years depending on the age differences and play styles of the cats. Over time, by teaching them how to interact with each other and engaging the more active cat with play and more mental and physical stimulation, they often relax and even like each other.

CAT RETURNING FROM THE VET

When a cat returns from a veterinary visit, the cat who stayed home may fail to recognize him. She may act defensively or aggressively and attack the returning cat. Often bringing cats to the veterinary clinic together prevents this issue.

APPROACH

When a cat returns home from the veterinarian, initially keep him separated from the other cat. Use a damp cloth or wet paper towel to remove the clinic scent from him. Wipe his paws and around and underneath his tail. Do this by massaging, stroking, or cuddling him with the moist towels or cloth. Repeat this four or five times over the course of a few hours.

Afterwards, rub the cat who stayed home with a dry towel or bandana and then rub and pet the cat who went to the vet with it. Do this three or four times over the course of a few hours. Repeat this process, but add your odor to the mix. To be extra cautious, dab some scented lotion or perfume on your hands and then stroke all of the cats.

Reintroduce the cats in a positive way so they can do something they enjoy when they see each other. This might be having them eat food together, playing with them with pole toys, or going out on the balcony.

If you know your cat is aggressive whenever you take another cat to the vet, request a short-term anxiolytic for the aggressor. Follow the above steps, but give this medication to the aggressor before reintroducing the cats. In addition, you may have to keep the cats separated for a longer period before reintroducing them, sometimes for a few days.

REDIRECTED AGGRESSION

When a cat is highly agitated, aroused, or startled by a sudden noise, scent, or an animal outside (usually another cat), she may redirect her aggression toward anyone nearby. This can be a person, dog, or another animal, but is usually another cat in the home (Figure 6.16).

Upon separating the cats, in many situations of redirected aggression, the aggressor is fine or 'back to normal' upon reintroduction. However, the victim is not. Often, the victim is so traumatized that when he sees the aggressor, he'll growl, hiss, or run. The aggressor can be extra sensitive to noises and quick movements when the victim is present. When this happens, a sudden movement, defensive response by the victim, or unexpected noise can make the aggressor attack the victim again.

Regression with redirected aggression is normal. This can take people by surprise, especially when they've made good progress with reintroducing the cats. Regression usually happens at night when cats are unsupervised, in tight corners when cats have to pass or walk by each other, or when both cats walk too closely in the same direction. It can also occur when a sudden, loud noise startles the aggressor and the victim is nearby. Because the negative association between cats is so severe – whether it is the location or context the fighting occurred in or the cats' proximity to each other – when the aggressor becomes heightened or aroused, she reattacks the victim.

Figure 6.16 Cats can become aggressive to other cats or animals in the home upon seeing another cat or animal outside or when there is a sudden loud noise. Binx & Floyd © Grace Yuen.

However, if you use the right approaches and go at each cat's individual pace, regression will happen less often. When reintroduced after separating them temporarily, the cats will show normal behavior quicker and bounce back faster after any altercations.

When there is redirected aggression, it's good to use the same approaches as if you were introducing the cats for the first time. (See 'Introducing Cats,' p. 115.)

With redirected aggression, both cats, especially the aggressor, can benefit from behavioral medication to make reintroductions easier. See your veterinarian for pharmacological medication.

APPROACH

If you suspect redirected aggression or have witnessed it, immediately separate the cats. Once the cats are separated, give them a few days to cool down.

When separating them, do not remove the victim from his preferred places in the home such as his favorite tower or sleeping with you at night. If you put the victim in a spare room for too long of a period while the aggressor has access to the prime areas of the home, the aggressor will likely be more aggressive when you reintroduce them. This can lead to territorial aggression exhibited by the aggressor.

If the redirected aggression is because of an outdoor cat, prevent your cat from looking out that location or block access to the area. You can purchase adhesive 'faux stained glass' for sliding doors and lower windows. Temporarily, remove perches from windows and move furniture away from any areas your cat may be able to see the outdoor cat.

Make note of 'hot spots' (where and when fighting occurred). Regression at these times and in those contexts is common. Initially, avoid those situations or locations, if possible.

Aggressive cats tend to be aggressive when the victim transitions from a higher surface (such as a windowsill) to the floor, makes a sudden noise or quick movement, or leaves the room. Think of these behaviors as triggers for the aggressor. It will be important to redirect and engage the aggressor at these times.

Before reintroduction, add more resources for the cats such as water bowls, litter pans, and window perches and place them in multiple locations. Since the cats no longer get along, they will not be able to share the same spaces. Add vertical territory and easy access to it for the victim so he can get away from the aggressor if pursued.

After a day or two of separation, allow the cats to explore all the rooms of the home separately. Make these explorations positive for the cats. Give treats, brush, or play with them in any areas where there may have been conflict. Whenever the aggressor sniffs locations or resources the victim has used, give her highly palatable food and praise her.

When the cats, separately, are calm in all rooms of the home, begin reintroductions. Watch your energy and relax around the cats. If you are nervous, frantic, or make a sudden noise, they will pair it with each other, not their behavior. If the cats show any signs of tension or there are subtle threats made by the aggressor to the victim, separate the cats for a few more days.

Think quality of exposure over quantity when reintroducing cats. Every exposure is a learning opportunity. It's better to have small, multiple exposures with positive experiences for the cats than one longer session that ends in a fight. You will go faster in the long run if you stay within each cat's comfort zone.

Make use of opportunities and situations to expose the cats in positive ways. If they like food or breakfast in the morning, reintroduce them at that time. If they are happiest in the office being on a desk or looking out the windows, introduce them in that location as you praise, brush, play with them, and give them treats.

Praise and give highly palatable food to both cats whenever they see each other. Redirect the aggressor's attention away from the victim using positive redirection or distractions.

Whenever the victim enters a room, makes a quick movement, exits the room, or gets down from furniture, praise and give treats to the aggressor. This prevents the aggressor from being startled or wanting to pursue the victim.

Make use of vertical surfaces when reintroducing cats. When the victim is relaxed on a higher platform, allow the aggressor to enter the room. Give treats to the victim, then to the aggressor.

Give treats to the victim to boost his confidence after any interactions with the aggressor. Reward him for entering rooms, showing initiative, and getting on higher surfaces.

Change the negative association of 'hot spots.' Pair positive things and activities for both cats, especially the aggressor, with the locations or contexts fighting occurred. It's important to change any negative associations the cats may have to a more positive one. You might play with them in that location or give them tuna juice (tuna water from the can) or catnip there. The good things don't have to be the same for each cat. Both cats should be exposed to each other at distances they are comfortable.

Keep the cats separated at night and when you are unable to supervise them.

When both cats are fine being together in the same rooms, meaning they can pass each other, get onto and off of furniture, exit rooms, and make quick movements or noises in each other's presence, without showing signs of aggression or fear, you can leave them together unsupervised.

If the aggressor is too unpredictable, shows any signs of serious aggression to the victim, or the victim is fearful and acts defensively around the aggressor, separate the cats, and after a period of separation, start again as if you were reintroducing them for the first time.

OVERGROOMING

Overgrooming can be common for some cats and certain cats seem to be predisposed to it.

Cats who excessively lick people, other cats, and sometimes surfaces may have a tendency to overgroom. Cats with medical issues such as cystitis can groom or lick themselves excessively on their abdominal or belly region. Ectoparasites such as fleas and mites can cause excessive grooming and itchiness, as can food allergies.

Food allergies or environmental allergens are broad categories. At this time, there is limited testing for cat allergies. So, the approach is usually providing special or novel diets and eliminating certain foods, and resolving symptoms by trial and error.

Normal grooming doesn't seem to affect serotonin levels, but excessive grooming can. Excessive grooming can increase serotonin levels. Cats may excessively groom themselves due to habit and repetition, so it can become a stereotypic behavior.

In most cases, overgrooming to the point of hair loss is medical, not behavioral. With severe overgrooming, cats usually target the lower abdomen or belly region and inner thighs, to the point where there is no more hair. The continuous biting and scratching at the area makes it red, inflamed, and irritated, which makes the cat orient to that area more.

If a cat is prohibited from grooming, chewing, or licking an area on her body, her fixation and desire will increase. If prevented from scratching at an area, she will scratch even more once permitted. Elizabethan collars, squirt bottles, constant interruptions, and trying to prevent your cat from grooming or scratching at an area that is irritated will only intensify her need and desire to do it.

In nearly all cases of overgrooming, going to the veterinarian and getting it medically treated with steroids, antihistamines, and anti-inflammatories help alleviate symptoms and break the pattern. This may have to be done for a period of time until the cat gets out of the habit. Often when medications are stopped, symptoms return. This also suggests there is a physical component to the behavior, not behavioral. In many cases with cats who overgroom, they are emotionally and behaviorally happy which further points to a medical issue.

If your cat is overgrooming due to behavior, you will likely see it in the following circumstances – boredom, after punitive corrections, in response to loud noises or conflict, or it occurs in specific situations, in certain locations, or at certain times.

Cats can lick themselves as a displacement behavior, so if there is stress in the environment alleviating it will be important. Overgrooming can occur when a cat is left alone too often or for too long, or when she is exposed to a continuous stressor she cannot get away from.

There is a syndrome called feline hyperesthesia which causes a cat's skin to twitch, especially along the back, hind end, and lower tail. The cat's skin may twitch or ripple in response to petting, as well. With feline hyperesthesia, the cat becomes agitated and often bites or attacks her hind end or tail to the point of self-mutilation.

There is a lack of medical research on feline hyperesthesia and it's poorly understood. Gabapentin seems to help cats with hyperesthesia, as do antihistamines, pain killers, steroids, and anti-inflammatories further indicating that this is physiological condition, unfortunately, for the cat and not a behavioral OCD (obsessive compulsive disorder). Although, it's often dismissed as behavioral, anyone who has had a cat suffer from it knows she is physically in pain and extremely uncomfortable.

APPROACH

Take your cat to the veterinarian. It's important to stop any itching or irritation. A change in diet may be necessary.

Add water bowls to rooms where your cat sleeps and socializes so she drinks more. This is important for her coat and health, and if she is well-hydrated, she should shed less.

Wipe your cat down daily or a few times a day with a damp paper towel or wash cloth to remove excess hair or any dust from her fur that may be irritating.

If the cat is bored or physically and mentally under-stimulated, enrich her environment. Add vertical surfaces and platforms for her to climb on. Incorporate interactive play sessions into her day to let her mimic hunting and stalking behavior, and provide her with affection, warmth, and a quiet environment.

Free-feed your cat or keep her well fed so she is not constantly famished or obsessed with food.

If there is inter-cat conflict, address the issues or dynamics between the cats. If the cat who overly grooms is a victim of aggression, it's important that she's no longer chased or harassed.

If your cat swats or hisses in response to touch, get her medically checked to see if there is an underlying physical issue. If nothing medically is found, it's best to treat it as a petting and handling issue (see 'Petting Aggression and Dislike of Handling,' p. 111).

BABIES, CHILDREN, AND CATS

In general, newborn infants do not cause much of a problem for animals since they are simply an extension of the mother. They may smell funny and make loud noises, but otherwise, they don't do much of anything. It is when the child turns toddler age that difficulties begin. The child walks and crawls on her own, may be encouraged to approach and interact with the cat, or may be left unsupervised with the cat.

A very young child or toddler can't read a cat's signals, nor does she have the motor control to know how gently to touch or stroke the cat. Since toddlers are on the floor, they are often at the same level as cats. This means if a cat swats a child, the child is swatted on her face or upper part of her body.

Since young male cats like to wrestle and kittens and young cats want to play, they can pounce on toddlers in surprise attacks as a means of playing. After they pounce on the toddler, the toddler falls and cries, and then they dash away. However, the cat was not intentionally trying to hurt or harm the child.

When children play with cats, they often throw or swing toys or wands toward the cat, instead of tossing or pulling them away from the cat. The child may think she's playing, but the cat wants nothing to do with it.

The body language of human toddlers and children can frighten animals, even friendly ones. Staring at an animal and direct approach can be viewed as a threat. Of course, this is what children and many adults do when they interact with cats.

Studies show that cats show little to no preference toward unfamiliar people regarding age or sex, but they do react strongly to how people behave toward them. Women and children tend to get down to the same level as cats. This is less intimidating than standing or looming over them as men are inclined to do. However, adult women waited for cats to approach them, whereas children, especially boys, tended to approach cats first. Boys followed cats who tried to walk away from them more often than girls did (Mertens & Turner, 1988).

To teach children how to be good with cats, management and supervision are key, and gentle guidance is vital.

APPROACH

Expose your cat to baby scents and sounds before your baby arrives. Wear baby lotion, baby powder, and wash clothes with any detergents you may use. Expose your cat to strollers, electronic mobile toys, rattles, and other baby items you might use when your baby arrives.

You can purchase and download audio of babies crying and baby noises. Play these softly at first while you feed your cat wet food or something extra delicious, or if your cat is play motivated, play with him as you play the recording.

Allow your cat to investigate the nursery. If you want to deter your cat from going into the crib, fill it with bottled water so it won't be as comfortable to rest there. Alternatively, you can allow your cat to sleep in the crib when your baby's not in it and close the door to the nursery or set up gates when your baby is sleeping, or simply remove him when he's in the crib and gently reposition him somewhere else. After a few times of repositioning, he'll likely choose another location. When the baby arrives, most cats lose interest in sleeping in the crib, at least when the baby is in it.

Allow your cat to sniff the baby and speak positively to him when he does. Brush your cat, play with him, give him special food, and cuddle him in the presence of your child. Don't alienate your cat from your baby by teaching him to avoid her. If you do that, when she wants to interact with him, he'll try to avoid her, run away from her, or potentially be aggressive.

When your baby becomes a toddler, she will spend most of her time on the floor. Increase vertical platforms and territory so that your cat has higher surfaces to climb on so he can get away from your child, but still be in the same room.

Your baby should not interfere or intrude on your cat while he is sleeping, eating, or using the litter pan. Install baby gates or use free-standing dog gates with cat doors to keep your toddler out of the litter pan area. Do not make it challenging for your cat to get to the litter box.

Install baby gates and free-standing dog gates with cat doors to manage interactions, especially if your child follows or runs toward your cat. By adding gates, you can allow your cat to have access to all the rooms of the home but prevent your child from chasing him.

Do not leave your toddler with your cat unsupervised.

Young children and toddlers can be taught to be gentle with animals. In child psychology, it's been shown that small infants and children mimic or copy the behavior of adults. They often practice or perform these behaviors when the adult isn't present. It is important to teach your baby from the start how to touch your cat gently and interact with your cat in a calm, quiet way.

Teach your baby to gently pet animals with an open palm, instead of grabbing at their fur. Human toddlers do not initially have the motor skills or coordination to keep their hands flat or palm open when they pet or stroke a cat, nor do they know how quickly or hard to touch him. Take your child's hand in yours, open her palm, and lightly stroke your baby's hand over the cat's fur and along his back. Do this very softly, so that your child learns how to touch your cat gently. Only do this when your cat is happy or relaxed.

Teach your child this nursery rhyme (I learned this as a little girl. It's easy for children to memorize):

> I know a little kitty. Her coat (or fur) is so warm.
> And if I don't hurt her, she'll do me no harm.
> So I won't pull her tail or drive her away.
> And kitty and I very gently will play.

Let your child give treats and food to your cat. The more treats your child gives to your cat, especially if your cat loves the treats she gives, the better the association will be between them.

INTRODUCING A DOG AND CAT TO EACH OTHER

Dogs and cats can coexist peacefully and happily together and great friendships can develop between them. Some dogs are very gentle with cats and prefer the company of cats over dogs. Other dogs need training and an adjustment period, while others need more management.

Some cats can be aggressive to dogs, seemingly without provocation. This tends to occur when a resident cat doesn't tolerate a new dog being brought into the home. Often these dogs are small, shy, or puppies.

Most young animals can adapt to each other quickly. Dogs and cats have very different body language which, initially, can be confusing when they try to interact with each other. For instance, when dogs greet each other and are friendly, they generally approach each other at an angle or from the side. A human equivalent might be an exaggerated European greeting or a do-si-do. Dogs look at each other and then look away again to prevent conflict. When two dogs stare at each other, they will usually fight or play. When someone directly stares at a dog, they are conveying to the dog that they want to engage or interact with him. For some dogs, this can be threatening and confrontational. When a dog looks or fixates on anyone or anything for longer than 2.5–3 seconds, he will likely want to approach or take action.

Cats, on the other hand, say hello to each other by nose to nose greetings. Cats will sit in middle of the foyer or hallway, on a stairwell, or the kitchen island and stare at the dog. Dogs who may want to engage will look at her staring at them as an invitation to interact or play. If the cat swats the dog without claws or as a warning, the dog might mistake it for play-fighting. When the cat lunges, hisses, or growls at the dog, it only confuses him (Figure 6.17).

Shy and cautious dogs will avoid eye contact with cats. They will turn or look away from cats if the cat growls or swats at them. These dogs tend to be excellent with cats since they ignore them. Because the cat feels safe, she engages more. Dogs who want to interact with cats have to learn to ignore them, even when the cat seemingly solicits them for attention.

In addition, cats like quiet, calm environments. Dogs are energetic, make quick sudden movements, and loud unexpected noises. This can be unsettling and startling for cats. If an animal or person suddenly enters a room, a dog's natural behavior is to bark or approach them. This can frighten the cat. When the cat transitions, jumps off of furniture, or leaves the room, the dog will want to follow. Of course, the cat finds this unpleasant and threatening.

Dogs are very social. If your dog is the only dog in the home, he'll likely follow you from room to room. Dogs do not like to be left alone. If your cat is used to being the center of attention or spending quality time with you such as sleeping with you at night or cuddling with you on the sofa, it's going to be hard for her to adjust to having your dog continuously present. Whenever you attempt to give your cat attention, your dog will want attention too. If your cat used to greet you at the door, your dog will beat her to it. Dogs don't particularly like sharing, and for that matter, neither do cats.

Figure 6.17 Dogs and cats have very different body language which can be confusing when they initially try to interact with each other. © Maksim Prochan.

Do not punish your dog or cat in the presence of each other. Although you may think you are correcting or scolding a behavior, they will think you are punishing them or that bad things happen to them when they are together. Neither the dog nor cat will pair your hostility or punishments with their behavior. Instead, they will pair it with each other.

Below are guidelines to introduce dogs and cats to each other. When you are not there to supervise, they should be separated until you know they will behave well together.

GENERAL APPROACH

Gates are excellent preventative tools. Let your dog and cat get used to each other's scent and sounds initially through a closed door. Then install extra-tall baby gates with doors and/or set up free-standing dogs gates in multiple rooms and locations. This way you can manage your dog and cat while you expose them to each other.

In addition to gates, make use of vertical surfaces for your cat so she can be exposed to your dog in a safe way. Do not bring your cat over to your dog or force any interactions.

If you wear a perfume or lotion, dab a little on your hands and then pet your animals or mist a little perfume or cologne on a cloth. Wipe yourself, your dog, and your cat with the cloth. Scent is important for animals. If your dog and cat smell similar to you and each other, they may get along or bond more quickly.

Feed your dog and cat extra yummy food and treats whenever they see each other as well as immediately afterwards. Your dog should only receive these special treats or goodies when he sees your cat. You can provide him with something delicious to chew on to keep him busy so his focus won't be entirely on the cat, or give him a lick mat filled with cream cheese or wet food, or a snuffle mat with bits of cheese tossed in.

Play with your cat, praise her, and give her highly palatable food or catnip whenever she sees your dog. Alternate between giving treats to your cat or playing with her and then giving treats to your dog. This establishes a positive association between them and gives them a reason to like each other.

Reward and encourage your dog to look away from your cat. Prevent him from chasing or following her through food and positive redirection. It will be vital to provide treats to your dog and positively engage him whenever your cat enters a room he's in, transitions or jumps onto and off of furniture, makes a quick movement, or walks away.

Reward your cat when she shows confidence, enters a room your dog is in, or jumps and climbs onto higher surfaces when your dog is present.

Through the use of baby gates, leashes, reward-based training, lots of goodies, and vocal control, you can prevent your dog from chasing your cat and teach him to interact with her in a positive way. When your dog is calmer and more mild-mannered around your cat, your cat will relax around your dog. Remember, if your cat hisses or growls at your dog, she feels threatened. Hissing is a defensive behavior, so when your cat hisses at your dog, she is fearful. It does not mean she will be aggressive.

TIPS FOR YOUR CAT

As a general rule, your cat should have access to all the areas of the home and be invited to these areas regularly.

To begin, designate a room or location that is entirely your cat's territory. Initially, she should have safe areas that your dog has little or no access to. Your cat should not have to run the gauntlet or be forced to pass your dog to access food or water or get to her litter pan.

Add vertical surfaces and shelving around the home for your cat to climb and rest on. Vertical platforms and higher furniture can distance her from the dog so she feels safer. Keep in mind a sudden jump to the floor can make a cat feel vulnerable and trigger a dog to chase, so add stable intermediate platforms and furniture so she can easily access these surfaces and walk down from them in multiple directions.

Position all your cat's resources – food and water bowls, litter pans, scratching posts, beds, perches, and toys – so that she cannot be unexpectedly cornered, trapped, or spooked by your dog and so that she has a great vantage point of the doorways when she uses them. If your cat has to pass through a cat flap to access her litter box or it's placed in the corner of the basement, she will feel uncomfortable using it.

Provide yummy treats, catnip, or play with your cat whenever she sees your dog. Make use of times when she is on higher surfaces, including the bed, to introduce your dog to her in a safe way. When your cat is less likely to run, your dog will be less likely to chase. It's important that your cat is not pursued or chased by your dog. Similarly, introduce your cat to your dog when your dog is resting, relaxed, or sleeping.

TIPS FOR YOUR DOG

Set up a baby gate or free-standing dog gates to keep your dog away from the litter box. Initially, your dog should not have access to your cat's litter boxes or her eating areas.

Use a front-connection body harness or flat leash and collar when introducing your dog to your cat. Keep the leash relaxed, regardless of its length. If you tighten up on the leash or jerk the leash to correct him, he will develop a negative association with your cat. If you pull back on him when he wants to see your cat, he will likely pull more. Either hold the leash so it's jiggly where it attaches to his collar or harness or let it drag behind him while you assess his behavior behind gates.

Exercise your dog before exposing him to your cat. Introduce him to your cat when he is tired, resting, or sleeping. If your dog has just run a few miles, taken a long hike, or chased a ball for an hour, he will have less energy and hopefully be calmer when meeting your cat. Do not introduce him to your cat when he has not had any exercise, or when he is restless, or has just woken up from a nap.

Feed your dog highly palatable food and meals whenever he sees your cat. This will make him look at you when your cat is present.

Expose your dog to your cat when she is on higher surfaces or on the bed. Try not to introduce him to your cat when she is on the floor. Being on the floor makes her vulnerable and encourages your dog to chase or follow her. Walk your dog into the room your cat is in or, if you are in the room with your cat, let your dog enter. Feed your dog wonderful goodies. Redirect his attention away from your cat in a positive way using praise, food, massage, or by petting him. Ideally, after you introduce him to your cat, lure him with food or a toy so he walks away from her or leaves the room.

It's important to teach your dog to ignore your cat or look away from her, especially when she enters a room he is in or makes a quick movement (Figure 6.18). You can do this by feeding your dog at the exact same time your cat moves, transitions (jumps onto or off of something), or leaves the room. If he stares at her, try to redirect him in

Figure 6.18 It's important to humanely teach and encourage your dog to look away from your cat, especially when your cat enters a room, suddenly moves, or walks away. © Danae Callister.

a positive way to a more appropriate behavior, such as chewing on a toy, coming to you, or lying down.

Positively train your dog when your cat is present. You might have to section off areas with gates when you teach him. Alternatively, train him when your cat is on higher surfaces so she can watch him from a distance. Teach him the basic body-positions, especially how to lie down on cue and how to settle. Lying down, especially in a relaxed down, staying, and making eye contact with you are far more effective than asking him to sit every time he sees your cat.

If your dog barks or lunges at your cat, walk him out of the room. Do not engage with him or praise him or make him follow any commands. Go back into the room your cat is in. When your dog calms down and is quiet, let him back into the room with your cat. Leave the leash on your dog so it drags behind him. This way you can easily remove him and prevent him from chasing your cat. You may have to remove him multiple times, but by removing him when he barks or lunges at your cat, you can teach him that he is banished whenever he barks or lunges, not that you don't like when he is around her.

When your dog can look at your cat and then look at you, or look, turn, or walk away from her, and when you can interrupt him if he begins to chase or follow her, you can expose them to each other without the need for any leashes or gates.

If your dog fixates on your cat, despite your best efforts, and you cannot get his attention away from her or he continues to pursue and chase her, it's best to keep them separated.

IF YOUR CAT IS AGGRESSIVE TO YOUR DOG

If your cat is aggressive to your dog, many of the same approaches apply as if your cat was aggressive to another cat (see 'Inter-cat Aggression,' p. 119).

Clip your cat's nails and keep them clipped. Use free-standing dog gates or install baby gates with doors so both your cat and dog can have access to all the rooms of the house, just not at the same time. Usually, when a cat is aggressive to a dog, it's when the space is too crowded or there is little distance between them. It's rare for a cat, even a very aggressive one, to leap over gates to get to a dog. Since most aggression stems from fear, when a cat is proactively aggressive, she really wants distance.

Keep them separated when you are not there to supervise.

Your cat can look at your dog, but it's important to prevent or interrupt her from staring at him. Give your cat treats, redirect her, open a window for her to look out of, or distract her with play if she begins to focus on your dog for too long. If interruptions and redirections don't work, reposition her. Be calm as you unemotionally pick her up and place her on a higher surface so she faces away from your dog. You may have to do this multiple times, mixing up where you place her down, but usually her focus will go elsewhere. If she still stares at your dog or tries to stalk him, remove her from the room temporarily. Try again at another time.

Praise your cat, give her yummy food, goodies, or anything else she may like whenever she sees your dog and immediately afterwards.

Introduce your cat to your dog when he is quiet, relaxed, or engaged in another activity such as chewing on a bone or licking a mat with yummy food, or let your cat be present when you massage your dog or when he is taking a nap.

Encourage your cat to be on higher surfaces so there is more distance between her and the dog. The floor is a vulnerable area for cats and where fights tend to happen. By getting your cat off the floor, she'll feel less threatened. She'll be further away from your dog and it will be easier for you to intervene when you need to. The further away your cat is from your dog, the easier it will be for her to accept him when he is active or if he makes a sudden movement.

Cats who are aggressive will be reactive to dogs when they move quickly, get excited, bark, or make a sudden noise such as when their nails click on the floor as they walk. If you have wood or tile floors, it's good to add carpets or runners to soften the noise when your dog walks and to give him more traction.

Tight quarters, corridors, and hallways, as well as when dogs and cats have to pass each other, will be points of confrontation. If your cat is too close to your dog or blocking an entrance or an exit, positively redirect her away from the area or pick her up and place her on a higher surface so your dog can get by.

If your cat's food motivated or likes catnip, give her something special after you reposition her, as well as whenever your dog moves, after you greet your dog, or after any other interactions you may have with your dog when she's present.

If your cat is highly reactive or aggressive, you may need medication to help her calm down. Speak with your veterinarian.

HOW TO RESOLVE A LITTER BOX PROBLEM

Behaviorally, there are many reasons why cats won't use a litter box. Most litter pan problems with cats are due to litter boxes being too small, too dirty, or in poor and vulnerable locations. In addition, many cats prefer to use separate boxes for defecation and urination. The style of the litter boxes (covered vs. uncovered) and types of litter may also be factors.

GET YOUR CAT MEDICALLY CHECKED

It's important to have your cat medically checked by a veterinarian if she is eliminating outside of the litter box. Feline lower urinary tract disease such as bladder stones, bladder infections or inflammation, urethral obstruction, as well as kidney disease, diarrhea, inflammatory bowel disease, and constipation are all reasons cats will avoid using the litter box.

Cats who have urinary tract infections will and can spray. Urethral blockage for a neutered male cat may cause him to adopt a standing posture to urinate. This may look like he's attempting to spray, but it is an emergency. The cat needs to be seen by a veterinarian.

UTIs (urinary tract infections) are very painful and can reoccur even after symptoms resolve, so they should be taken seriously. If a cat experiences any pain or discomfort when using the litter box, whether for urination or defecation, he will avoid it or the location it is in and go elsewhere.

Veterinary diagnoses are generally determined through one or all of the following: bloodwork, urinalysis, urine culture, bladder X-rays, cystoscopy, and ultrasound.

DO NOT PUNISH YOUR CAT!

Do not yell at or punish your cat. Your cat is not urinating on your clothes to spite you or make you angry. Cats do not know how to handle stress well, so purposely getting you upset is not your cat's intention. Never rub your cat's nose in urine or feces. (Some still recommend this for dogs. Don't do it to dogs either!) This will make your cat fearful to urinate or defecate in front of you and will cause more anxiety. Stress can weaken a cat's immune system and stressed cats can develop urinary tract infections. In addition, if your cat is hesitant to eliminate, he will try to avoid or delay eliminating altogether, which is unhealthy. A buildup of urine in the bladder can lead to infection which can then affect the cat's kidneys.

If you catch your cat eliminating in an unwanted spot, unemotionally pick him up and place him in the litter box. Do not try to hold or confine him to the box. More than likely, when you place him in the litter box, he will immediately hop out since he is avoiding it to begin with. If he uses the litter box, praise him. You can certainly give him treats afterwards but giving treats to a cat for using the litter box will not resolve a litter box problem.

To resolve a litter box issue, it is important to get to the root of the problem and try not to make changes piecemeal. If there are multiple cats and one or more cats are defecating or urinating outside of the litter box, it's important to look at their relationships to see if there is any underlying conflict, in addition to evaluating the litter box arrangement or set up itself. (See 'Inter-Cat Aggression and Conflict,' p. 119.)

If your cat is scratching the sides of the box, the walls, or floor outside of the box, not digging or covering, balancing on the edge of the box, or going near the box but not into it, the litter pan is likely too small, too dirty, or both. A dirty litter pan is like an unflushed toilet – no one wants to use it unless there are no other options and one is desperate. When there are multiple cats, if one cat urinates or defecates in the pan, another cat has to step in soiled litter. It might just be easier and cleaner for the cat to go on the carpet.

Everyone likes an upgrade and cats are very good at showing their preferences. If your cat likes the changes you make, he will let you know. So, it's important to carefully observe his behavior.

CLEANLINESS

Ideally, clean the litter box by scooping the urine and feces multiple times a day or after each use, just as you would flush a toilet. Minimally, once in the morning, once in the afternoon or when you return from school or work, and once before you go to bed.

Boxes should be washed with hot soap and water every few weeks or the urine smell soaks into the plastic. This is especially true if fresh litter isn't added to the box as you scoop. Do not use strong smelling cleaners.

If cats in a household prefer one location or litter box, you will have to scoop and clean that litter box more frequently or add additional boxes.

If you can smell the litter boxes when you walk into a room, they are dirty.

SIZE

Most store-bought litter pans tend to be too small for adult cats. This is particularly true for larger cats and long-haired cats. The smaller the box, the dirtier it gets. By adding a hood or lid to the box, it can make the box even smaller and more restrictive.

The general rule for size is the larger the better. The minimum size should be the length of your cat, plus a few inches, and if your cat turns around, all four paws can remain in the box (Figure 6.19). Often, cats have to step outside of the box and/or

Figure 6.19 The best litter boxes for cats are open, shallow under the bed sweater or storage boxes. The box (pictured) should be the minimum size for an average-small cat.

bump into the sides or walls of it to turn around. This prevents them from moving or scratching as they normally would. If an open litter box is flush to a wall, it restricts the cat's movement and access to it.

For any litter boxes, open or closed, your cat should be able to scratch, cover, and turn around without bumping into anything. Sometimes, this can be as simple as placing an open litter box 7″–10″ away from the wall.

LITTER

Plastic liners can deter cats from scratching the litter and using the litter box. So, if you have plastic liners, remove them.

Cats prefer low-dust, unscented, low-tracking, clay or fine, sandy litters. Clumping is best to remove urine and feces. When litter is non-clumping, the urine stays in the box and makes it smelly.

Feline pine, pellets, tea leaves, cedar chips, and crystals, although popular, are not preferred substrates for cats.

The depth of the litter should be deep enough for your cat to dig and cover without scratching the plastic bottom or sides of the box. If there is not enough litter in the pan (you will be able to see the plastic bottom), the box gets soiled quickly and urine and feces seep into the plastic. It's good to top boxes with fresh litter after scooping to ensure they remain clean.

STYLE OF BOX

There are two common styles for litter boxes – open vs. covered. The best litter boxes are open, shallow sweater boxes or shallow storage containers. Cats have a better view and more room to turn around in an open box.

Sometimes, cats raise their hind end or stand up as they urinate, or they start in a squatting position and finish in a standing position. I call these cats 'high risers.' This usually occurs when the litter box is flush to a wall or vertical surface. It can also happen when high-sided boxes are not large enough. For cats who urinate this way, provide an open, shallow box, but don't place it too close to the wall. If you want to provide a box with high sides, make sure the box is large enough for the cat to turn around in without bumping into the walls. The cat should easily be able to enter from the front and, ideally, should be able to see over the sides of the box.

If you prefer a covered box, make sure the box is as large as possible and that the entry is at the front of the box, not on top of it. A covered pan can make a cat feel vulnerable since he can be trapped or cornered when using it (Figure 6.20). The scent is usually stronger in covered boxes as well, which makes it better for us, but less pleasant for the cat.

Do not use high-sided tubs, domes, and top entry litter pans. Although these are popular, they are not preferred by cats (Figure 6.21). I also don't recommend 'toilet training' your cat. This goes against the nature of the cat since cats like to scratch at litter, as well as urinate and defecate in different locations. Cats do not eliminate in water, and it's awkward or difficult for the cat to maneuver on a toilet.

Figure 6.20 A covered litter box can make a cat feel trapped or vulnerable using it. © Heather Rushton.

NUMBER OF LITTER PANS

Cats like more than one box. Many cats prefer to urinate and defecate in separate litter pans or locations. You may need more than one box for one cat. Stool is dirtier than urine, so oftentimes a cat will urinate more than once in box, but if there is feces in it, he will go elsewhere. This becomes especially problematic when multiple cats share litter boxes.

If you have more than one cat or a cat who is particular about the litter box, it's important to have multiple boxes.

LOCATIONS

Litter boxes should not be placed next to food, water, or adjacent to scratching posts or pads. Cats prefer not to eliminate in their social areas.

Figure 6.21 Avoid high-sided tubs, domes, and top entry litter boxes. © Thorsten Nilson.

Litter boxes should be in relatively private locations, but easy access for the cat. They should be positioned or angled so that when the cat uses them, he can see the doors or entries and exits of the room. Avoid keeping litter pans in the basement. Basements tend to be too far away, especially if the home has multiple floors. Litter pans in basements tend to remain dirtier because they are cleaned less often. Fearful, timid, and elderly cats will not want to take a hike or journey to get to the litter box. Ideally, there should be a litter box for the cat on each floor, or minimally he should only have to walk up or down one floor to access it.

It's a good idea to place litter boxes in a variety of locations, especially if you have multiple cats. Different cats may have separate and preferred locations. When all litter boxes are placed side by side and in the same location, it can be awkward for the cat to use them. Instead of positioning boxes flush next to each other, put some distance between them. If you have two small boxes side by side, it's probably better to get one large box. Likewise, cats should be able to access boxes without having to walk through or past other cats. They shouldn't have to walk through one litter box to get to another.

HOW TO CLEAN STAINS OR SOILED LOCATIONS

Soiled areas should be cleaned with enzymatic cleaners and non-ammonia-based products designed to eliminate or neutralize the odors of cat urine and feces. Do not use ammonia-based cleaners since ammonia is a component of urine. The use of vinegar alone will not deter further urination. Likewise, don't use simple carpet deodorizers or basic cleaners such as Lysol or Febreze. They may temporarily mask the odor for us, but not eliminate it for the cat.

It's important to locate all soiled areas and clean them thoroughly. Otherwise, the smell of urine will attract the cat to urinate there again. To locate soiled areas, use a black light or crawl on your hands and knees and take the 'sniff' test. Likely, you will have to do both.

If you use a black light, urine should fluoresce or glow (Take note that not everything that fluoresces is urine). Black lights don't work perfectly on all surfaces. When sniffing for stained areas, find spots that do *not* smell like the surrounding dust, wood, plaster, or paint. If you smell urine, the cat definitely smells it too. When you can no longer tell what you're smelling, take a break for ten minutes or sniff some coffee beans. This will refresh your brain so you can start sniffing the room again. Mark any old stains you find so you remember where they are. Then, soak those areas with hot water. After 5–10 minutes, absorb the excess water with cloth or paper towels. Afterwards, saturate the area with the 'enzymatic' cleaner you choose. Let the cleaner soak into the floor or carpet and then air dry.

If your cat has been urinating on walls or baseboards, clean and neutralize the area as best as you can. Then repaint the baseboards or that section of the wall. This often resolves the issue if the cat has been urinating there due to the smell.

CHANGING ASSOCIATIONS – THERE'S NO NEED FOR TIN FOIL!

It's important to change the cat's association with previously soiled areas to make it less appealing for him to eliminate there. Place water bowls, cat trees, window perches, edible plants, or scratching posts in the area. Feed your cat treats or brush him there. Most cats will prefer not to eliminate in their resting, eating, sleeping, or drinking areas.

If your cat is going on soft surfaces or furniture such as your bed, sofa, or armchair, get 'chux' pads (bed pads for people) available online or at drugstores or unscented housetraining pads for dogs. Using chux or unscented housetraining pads is a simple solution to preserve your clothes, bedding, and furniture while you determine the source of the problem. Line the sofa or bedding with the pads and then cover the pads with a waterproof blanket (also available online). This way you can use the furniture to cuddle, nap, and play with your cat, yet still protect it.

Play is an excellent way to prevent problem urination in a particular area. Play in the locations you don't want your cat to eliminate in. Cats do not want to soil their play area or hunting ground. End play by letting your cat catch the toy and then feed him treats in that location.

If the location doesn't allow you to change the association for your cat by playing with him, giving him treats, or placing relevant cat resources there, use a Feliway diffuser or spray. This can prevent problem urination. Don't plug in or spray Feliway near or next to litter boxes.

If you feel that your cat is anxious and could benefit from a behavioral medication, contact your veterinarian.

REFERENCE

Mertens, C. & Turner, D. C. (1988). Experimental analysis of human-cat interactions during first encounters. *Anthrozoös*, 2, 83–97.

7

GERIATRIC, AGING, AND UNWELL CATS

INTRODUCTION

Cats can live into their late teens and early twenties, so often what we think of as an old cat (7, 8, or 9 years old) isn't particular old, nor would I consider them 'geriatric.' However, as cats age, they become frailer. Most cats over 12 have arthritis. Medical issues common in older cats, especially in their mid- to late teens, are hearing and vision loss, hyperthyroidism, periodontal disease, and kidney and liver disease. Just like people, cats urinate more frequently when they age as well. Cognitive dysfunction (Alzheimer's) also affects cats. Cognitive dysfunction and arthritis in cats have only recently been recognized by the veterinary profession.

For many cats, the first sign of a health issue is a change in behavior. This is especially true for older cats. Excessively loud vocalization is common for elderly cats. This is usually associated with hearing loss, but it can be associated with hyperthyroidism. Increased pacing, restlessness, and night-walking can be signs of cognitive dysfunction. Hunger, thirst, or pain can cause increased vocalization and restlessness. Pain and illness will cause decreased affection or a cat to isolate herself. Signs your cat is in pain or cold will be decreased grooming and having a huddled or crouched body position (the bread loaf kitty).

No matter how upset or tired you are, do not yell at your cat. This will only damage your relationship with her and make her upset. It will not change the underlying cause of the behavior or the behavior itself. Your cat will simply become anxious and frightened of you.

If you see any behavioral or physical changes in your senior cat, it's important to take her to the veterinarian to rule out medical issues.

ENVIRONMENTAL CONSIDERATIONS

It can be painful for older cats to jump onto and off of furniture. If you watch your cat, you may see she hesitates getting down from higher surfaces. Even the distance from a windowsill to the floor can be too long of a leap. For this reason, older cats can spend much of their time on the floor. Add stable intermediate platforms and ramps so your cat can easily get to preferred spots and higher surfaces (Figure 7.1). Plush rugs or yoga mats can be positioned under windowsills and furniture.

DOI: 10.1201/9781003351801-7

Figure 7.1 Add stable intermediate steps, platforms, and ramps for older cats. © Lens on Focus_ Dreamstime.

Provide plenty of water bowls and place them in obvious locations. The best places for water bowls are where your cat spends the most of her time sleeping or napping. Keep water away from the litter pan and clean and refresh the water bowls daily.

Make sure all litter pans are easily accessible for your cat. It shouldn't be a journey for her to get to them. Sometimes, high-sided boxes are necessary for elderly cats so they can balance and lean against the walls as they use them. If you get a high-sided box, make sure the front entry is low so she can easily enter and exit from it. Since older cats tend to urinate more often, you will have to scoop the boxes more regularly and add new cat litter to the pan to keep it fresh.

Provide heated beds and warm, comfy blankets for your cat to lie down on. Older and arthritic cats get cold easily. Colder temperatures can increase pain from arthritis. It's uncomfortable for elderly and underweight cats to lie on cold, hard surfaces so make sure there is always cushioned padding underneath them.

Since vision loss is common for older cats, add better lighting throughout the home. Add nightlights to hallways, along stairwells, near litter pans, and food and water bowls. Add sound cues to areas your cat may need to access. A ticking clock near the litter pan or a running fountain near her perch can give her auditory cues to guide her to preferred resources.

When playing with an older cat with poor eyesight, play with toys that make noise. Use the metal clip at the end of a string (with no attachment) so she can hear it as it moves along the floor. Rustle a pole toy in a paper bag or move it along and around the sides of a cardboard box. Even sliding a piece of paper on the floor can be entertaining for a cat. Nylon tunnels can be good for elderly cats because they make a crinkly sound when cats enter them. They also make noise when you move a wand or pole toy under or around them.

If your cat suffers from hearing and vision loss, use scent markers to orient her to resources she needs to access, as well as different floor substrates and carpet textures for her to walk on. For instance, a plush rug can lead to her water bowl. A Berber mat can lead to her litter pan, or a yoga mat can lead to her food bowl. Avoid changing the placement of resources. Try to keep them in the same general areas or locations for her.

It's important for cats to eat. As cats age, many have their teeth removed. Provide your elderly cat with more wet and moist food, cooked meats, and tuna water. These can add variety to her diet and provide her with needed liquid. Do not restrict an elderly cat's food. If your cat is on a special diet, she should have full access to the food whenever she wants it.

HANDLING

Use a soft brush and moist paper towel to groom and clean your cat regularly. Gently clean her face and wipe her eyes. As cats get older, it's harder for them to clean and groom themselves.

Keep her nails trimmed. Due to arthritis and frailty, scratching behavior and activity decrease as the cat ages. This can cause her nails to grow longer than they normally would. If her nails are too long, ingrown toenails can develop which are painful and it can make it uncomfortable for her to walk.

Any handling for older cats can be painful as well, so it's important to be gentle and supportive when holding, touching, and carrying them. When you pick up and carry an older cat, support her chest and pelvis and try to keep her spine in line with the rest of her body. Always place older cats down gently and steady them as you do.

If your cat seems in pain or needs to be handled regularly for medical procedures, see your veterinarian and request she be given ample pain medication.

HANDICAPPED CATS

BLIND CATS

If your cat is blind, she may require extra care but she can live a long and happy life. When you have a blind cat, it's important to make use of the senses she does have. A cat's hearing is much more developed than ours. Cats can hear sounds above and below our hearing range and their ability to locate the sources of sound is highly advanced. They also have an excellent sense of smell (Figure 7.2).

Use sound cues and verbal signals to get your cat's attention and to let her know when you are entering a room. Tap a fork or spoon on the side of a plate to signal it's mealtime. Fountains can be used for water bowls. Fluffing up sheets, pillows, and patting the bedding can indicate bedtime. Opening blinds and curtains can indicate there is a sunny spot by the window and hanging different sounding bells on different doors can let her know what rooms you are entering and exiting from.

Substrates and textures can also guide blind cats. Different styles and textures of flooring, rugs, and furniture fabric can direct a cat to her favorite spots and let her know where resources are. A shaggy carpet runner can lead up to the sofa or a yoga mat can lead up to a windowsill or edible plant.

Scent can also let blind cats know where things are. Warm up food before serving it or before giving her treats blow at them in her direction. Catnip toys, edible plants, and different furniture fabrics and surfaces have different odors too. Large stuffed toys filled with catnip can be placed in walkways and where she socializes so she can bite

Figure 7.2 Blind cats can have long and happy lives. They have excellent hearing and sense of smell. Lewis © Christine McKinnon.

and bunny kick them. Catnip can be scattered on scratching posts and scratch mats. Cats who aren't responsive to catnip may be attracted to silvervine.

A blind cat will be aware of things in her environment we don't pay attention to. Blind cats are sensitive to temperature and, like all cats, are attracted to heat. If you can, keep rooms at different temperatures or add heated beds and pads to her favorite spots.

Talk to blind cats regularly. Give them a sweet verbal heads up when you enter a room or lean in to pet them, and verbally cue when you are going to pick them up, cuddle, or kiss them. This lets them know where you are and also indicates your intentions.

Play with blind cats using toys that make noise, and use boxes and tunnels made of different materials. Most cat tunnels are made of nylon and have a crinkly sound to them, while cardboard boxes are smooth and have a particular odor. Crumpled paper balls, a clip at the end of a string that you slide on the floor, or moving a toy behind or under a paper bag can let a blind cat know where the toy is. When providing boxes and tunnels, be sure to keep them in the same general locations or arrangement.

It's very important to look out for a blind cat's safety. Block off openings to stairwells, banisters, balconies, and ledges, as well as access to the fireplace. Keep toilet lids closed. Add fencing to decks. Dull or cover sharp corners and edges of furniture with padding and close doors to closets or other areas that may be confusing to her.

Provide intermediate platforms and ramps so blind cats can easily access furniture and higher surfaces. Make sure there are sides to all ramps and platforms and that they are large and wide enough so that if she misses a step, she doesn't fall. Place soft padding, yoga mats, and cushioning under any surfaces or platforms she may use, including windowsills, so that if she does fall, she doesn't hurt herself. If you have a porch or balcony, screen it in, or if you are able to, build a catio or outdoor enclosure for her so she can enjoy the outdoors while staying safe.

Blind cats rely on memory. One of the most important things you can do for a blind cat is to keep all resources in the same general locations or arrangements for her, even if you have to move. When you are in a new environment, direct her to her resources by guiding her with your hands and walking alongside her on your hands and knees. Use touch, scent, sound, and any other signals she is familiar with to show her where her resources are.

Blind cats frequently walk and lean against walls, fencing, and familiar objects to orient themselves. Keep this in mind when positioning resources for a blind cat and when placing furniture.

DEAF CATS

Deafness can be congenital (many blue-eyed white cats, especially females, are born deaf) or related to age or injury. Most cats lose their hearing as they age. Side effects of anesthesia can also cause deafness in cats.

Deaf cats can cry out more and are louder than cats who can hear. This is especially true for elderly cats and when a cat suffers from sudden hearing loss. If born deaf, some cats may not cry at all. Because deaf cats can't hear you, they can easily be startled or spooked when you touch them, and they may seem unresponsive when you talk to them or call them.

If you have a deaf cat, make use of their vision, as well as other sensory cues they have such as smell, vibration, warmth, and touch. Since humans are so verbally oriented, it can be hard for us to think of communicating in visual, non-auditory, or tactile ways. Once you get their attention, deaf cats can learn to respond to hand signals.

If your cat is lying on a sofa or bed, press on the cushions or bedding next to her to get her attention. If she is sitting on the floor, tap on it lightly so she orients in your direction. Pair your touch with treats so you don't surprise her. When she is happy or relaxed, from behind her, lightly tap on her shoulder. When she turns in your direction, give her a treat. Tap on her right shoulder to signal her to look right and her left shoulder to teach her to look left.

For meal times, warm food before serving it or blow at it in the cat's direction. To get her attention, flicker a light. Cue playtime, brushing her, cuddling, and coming to you by showing her visual signals beforehand. With repetition, she will associate your hand signals with the behavior or certain activities.

If you have a larger home or multiple animals, it can be helpful to put a belled collar on your cat to keep track of her whereabouts, especially when she's in another room or out of sight.

DEAF-BLIND CATS

When a cat is deaf and blind, it can be challenging for them to navigate their world or to think you can give them a quality of life. However, deaf-blind cats can feel, taste, and smell. The feel of textures, surfaces, and substrates, keeping resources in the same order or general locations, and using scent, temperature, vibrations, and air movement are all ways you can communicate with deaf-blind cats.

Instead of visual, auditory, and verbal communication, think tactile signals, vibrations, and odor.

A deaf and blind cat may feel your air movements and vibrations as you approach, walk away, leave the room, or open and close a door. The cat's sense of smell will direct him too. Sisal rope, cotton cords, furniture, and carpet fabric all have different odors and textures. Use these to lead and guide a deaf-blind cat to resources and certain locations.

Blow at treats in his direction when you want to feed him and warm up food before serving it. The smell of coffee, tea, and food, as well as vibrations from the refrigerator, can let a deaf-blind cat know you are in the kitchen or where it is. If you have a favorite perfume or lotion, use it for a particular area or room such as a bedroom or bathroom. Provide catnip scented toys and edible cat safe plants for deaf-blind cats to chew on and scatter catnip on scratching posts and scratching pads. If a cat isn't receptive to catnip, try silvervine instead.

Blind-deaf cats can feel differences in temperature and are attracted to warmth. Add heating pads or heated beds to sleeping areas, or if you have the option, keep rooms at different temperatures. You can open a window or door to the outside (be sure they are both screened in) so he can lie in the sun or get some fresh air.

Your style and manners of touch will also be cues for a deaf-blind cat. You can teach him to turn right, left, or to stay where he is. When he is sitting, lying down, or resting, press gently on the bedding or pillow next to him or tap on the floor to get his attention. When he is affectionate or when he knows you're there, touch him on the shoulder and give him a treat. Touch his right shoulder so he turns right and his left shoulder so he turns left. To give him cues to stay settled, after petting or engaging with him, rest your hand on him gently and keep it still. It's important that whatever signal you use, to keep it simple and be consistent. Over time, you can add to them or build on the skills he already knows.

Try not to move the arrangements or placement of food, water bowls, and litter pans. Think directionally such as right and left. Blind-deaf cats use their memory to access resources and to guide themselves. The moment something changes, they will be confused. If a deaf-blind cat is disoriented, they will remain still. If they are social or bonded to someone, they likely will cry out for help.

If you move, try to keep the cat's resources in the same general orientation. When in a new environment or when introducing new things to a deaf-blind cat, walk on your hands and knees alongside him and guide him with your hands and any other sensory cues he is familiar with.

Lastly, it is pivotal to protect a deaf-blind cat and keep him safe. He should wear a belled collar so you can keep track of where he is. Add large intermediate ramps and steps up to any furniture or platforms he might use and place soft mats and padding underneath them, so that if he falls, he doesn't hurt himself. Block off access to unsafe areas such as fireplaces and balconies, and add stable walls and sides to ramps, ledges, walkways, or stairs he may use, so that he doesn't fall if he misses a step.

Blind and deaf-blind cats frequently walk and lean against walls, fencing, and familiar objects to orient themselves. Keep this in mind when placing furniture and all of the cat's resources and belongings.

CEREBELLAR HYPOPLASIA

You may have seen cats with cerebellar hypoplasia. They are uncoordinated, wobbly, zig-zag when they walk (or seemingly go in multiple directions at once), and have head tremors. CH cats are born to mothers who suffered from Feline Distemper, Feline Infectious Enteritis, or Feline Panleukopenia during pregnancy. The virus affects the area of the cerebellum responsible for fine motor control. Cats with cerebellar hypoplasia can adapt quite well. They have the same lifespan as normal cats, although they may not grow as large. They can have sight problems and have slower development.

Because cats with CH are poorly coordinated and can't walk in a straight line, it's important to keep them safe. CH cats want to play, climb, and scratch just like any other cat. However, they can tip and fall over easily. They often lean against walls and the sides of furniture for stability.

It's important to provide them with wide and stable intermediate steps, ramps, and platforms so they can easily access furniture and higher surfaces. The steps or ramps should be carpeted. Adding protective sides to ramps and stairwells is necessary so they don't tumble if they miss a step or lose their balance. There shouldn't be any gaps between stairs. Soft padding and flooring, cushions, or yoga mats should be placed under windowsills, furniture, or any vertical platforms they use, so if they do fall, they don't hurt themselves.

All resources, including litter pans, should be easily accessible. Since cats with CH lean against walls to balance themselves, litter boxes should be large and high sided, but with a low entry. CH cats squat when they urinate, but since they have poor coordination, they can tip over easily when they try to cover, scratch at the litter, or turn around. If they miss or dribble over the edges of the litter box, place pee pads underneath and around it.

Although they try, cats with cerebellar hypoplasia cannot clean themselves as easily as other cats. You will have to wipe their faces after they eat, clean under their tails, pet and stroke them regularly with moist paper towels, and brush them frequently to keep them well-groomed. They can be messy when they eat, so plates and bowls should be wide and stable. Mats can be placed underneath the bowls to catch any water that spills or food that's dropped.

CH cats can acclimate to a harness and leash if you have a quiet area to take them outside. Enclosed porches and catios are great for CH cats as they can provide them with outdoor time and enrichment. CH cats have the same interests and behaviors as any other cat, so except for their poor coordination and wobbliness, they can have an enjoyable and fulfilling life.

THE THREE F'S: FELINE IMMUNODEFICIENCY VIRUS (FIV), FELINE LEUKEMIA (FELV), AND FELINE INFECTIOUS PERITONITIS (FIP)

FIV

Feline immunodeficiency virus (FIV) is similar to the human immunodeficiency virus (HIV), but its symptoms are milder in cats. It is feline specific, so only cats can get FIV. Cats with FIV have weaker immune systems, so they are more susceptible to infections, such as gum infections and gingivitis, and healing can be more difficult for them. FIV+ cats often need their teeth removed which resolves any mouth pain or discomfort that may develop. In addition, FIV+ cats are more susceptible to lymphoma.

FIV is not particularly contagious to other cats. The virus does not survive outside of the body so it is not transmitted through sharing spaces, food and water dishes, or mutual grooming. Transmission is through deep bite wounds or semen, so intact males are more likely to be infected due to their roaming and fighting tendencies. Spaying and neutering cats reduces FIV transmission. FIV+ tom cats with symptoms, once neutered, can often return to normal health with proper care and diet.

Contrary to what has been believed and promoted, neutered FIV+ cats and FIV– cats can live together. It is extremely unlikely an FIV– cat will get the virus from an FIV+ cat.

Over 50% of FIV+ cats have the same lifespan as FIV– cats and show no clinical symptoms. Some cats can clear the virus with good care and nutrition.

The long accepted recommendation by the veterinary profession was to euthanize FIV+ cats. Most animal shelters and animal control facilities still kill FIV+ cats upon intake or, if they accept them, they house them with feline leukemia cats. However, these two diseases are very different. The advice of killing cats with FIV should no longer be recommended.

Unfortunately, a common test for FIV, the ELISA or SNAP test, detects antibodies for the virus, but not the virus itself. Kittens under six months of age can test falsely positive since they receive antibodies from their mother and it can take up to six months for their mother's antibodies to disappear.

There is a vaccine for FIV, but it has little efficacy, and there is really no reason for it. Be aware that once a cat is vaccinated for FIV, he will test positive for it on any ELISA or SNAP test.

Personally, I've had multiple FIV+ cats over the years. They have always lived with my non-FIV cats. None of the FIV– cats ever contracted the disease.

FELV

Feline leukemia is another disease that is common in cats. FeLV is spread through bodily secretions such as saliva. It can be passed from one cat to another through grooming, as well as from mothers to kittens, and through bite wounds. Feline leukemia attacks the bone marrow and cats with FeLV have compromised immune systems so are more prone to upper respiratory infections, dental disease, and anemia. Many cats have no noticeable or very mild symptoms since the virus is often dormant until there is a physiological stressor.

Since the virus does not live long outside the body, FeLV is not easily spread via clothing or hands. It is not spread through waste so cats cannot get FeLV by sharing litter boxes. Though transmission is unlikely, for safety, food and water bowls shouldn't be shared. FeLV+ cats can live in the same home as FeLV negative cats, as long as they are provided separate rooms.

At this time, there is no treatment for FeLV other than supportive care and treating and monitoring secondary infections. FeLV cats can live happily and be healthy, but they do have shorter lifespans.

The ELISA test, which is a standard test for FeLV, tests for antibodies to the virus, not the virus itself. Cats who tests positively for FeLV on the ELISA or SNAP test may not have FeLV and may never succumb to the disease. An IFA, immunofluorescence assay, is more conclusive.

FIP

Most cats naturally carry the feline enteric coronavirus. It is usually harmless and lives in cells of the cat's intestinal tract. It is shed through feces. A mutation of the feline enteric coronavirus causes a disease called feline infectious peritonitis. There are two forms of the disease, wet and dry, that manifest different symptoms. The virus is contagious to other cats and can stay active on surfaces outside of the body.

FIP affects all felines, including wild cats, especially kept in captivity. There is a high prevalence of FIP in captive Cheetahs. This is likely due to poor immunity from stress and confinement.

FIP can be transmitted from the mother to her offspring. Kittens and young cats under three years old are more vulnerable since their immune systems are not very strong. Once a cat's immune system is stronger and the cat becomes a full adult, the prevalence of FIP decreases. Adult cats exposed to FIP can recover from it and then will have antibodies for it.

The ELISA and IFA (Immunofluorescence Assay) tests will show the presence of coronavirus antibodies, but that only means that a cat has been exposed to FIP. Many cats who develop antibodies to a virus don't develop the disease, so a cat's level of antibodies is not a predictor of whether or not the cat has the illness or virus.

There is an intranasal vaccine to prevent FIP, but it has poor efficacy. Keep in mind, once a cat is vaccinated, she will test positive for FIP on both the ELISA and IFA tests.

Souki, my very first kitten I adopted in college, had FIP. The veterinarian had to do her own research to find the cause of his symptoms since FIP was not very known at that time and there was little written about it. Fortunately, she was passionate and determined to find the cause of his symptoms. Unfortunately, I was not there at the time of his euthanasia and there were no cell phones back then.

INTERSTITIAL CYSTITIS (INFLAMMATION OF THE BLADDER)

In medicine, idiopathic means the cause is not yet known. Idiopathic cystitis is inflammation of the bladder and goes by a few different names – feline lower urinary tract disease (FLUTD), feline interstitial cystitis (FIC), and FUS (feline urological syndrome). FUS is caused by crystals in the urinary tract. When urine is more concentrated, it is more likely to form crystals which can form stones. Urethral blockage or urinary obstruction is common in neutered male cats and is an emergency that needs to be immediately treated.

If your cat tries to urinate frequently, but little urine comes out, seems in pain or uncomfortable, licks at his genital region or abdominal area, or suddenly stops using the litter box, it's important to get him checked by a veterinarian.

Although there can be a number of causes, stress can exacerbate and increase the risk of feline cystitis. Cats who spend most of their time indoors and lack environmental enrichment or physical exercise are more prone to suffer from it. Fearful and stressed cats, cats in shelters or where there is overcrowding, as well as cats who live in unsanitary conditions are also susceptible. Most cases occur in cats kept indoors. Lack of water or minimal water intake and diet are also factors.

It's important for cats who are prone to urinary issues and bladder inflammation to have access to fresh, clean water. Add fountains and large water bowls near the cat's sleeping and resting areas. Let him drink from the sink or bathtub and keep water bowls away from the food bowl and litter pans.

Even though interstitial cystitis or feline lower urinary tract disease is a medical issue and the reasons for it may not be entirely known, environmental enrichment for the cat and limiting his stress levels are important. A clean, quiet environment, interactive play sessions, warm safe places to rest, the ability to climb and access vertical surfaces, as well as proper placement of food, water bowls, and litter boxes contribute to the cat's overall well-being.

HERPES VIRUS (CONJUNCTIVITIS)

Feline herpesvirus 1 (FHV-1) or feline rhinotracheitis virus is the most common upper respiratory infection in cats. It's sometimes referred to as the 'cat flu.' It causes conjunctivitis and severe head cold or flu-like symptoms in cats and kittens. Kittens and cats with compromised immune systems are at higher risk. One major symptom besides a snotty, congested nose is inflamed, red, or swollen eyes. The eyes produce discharge, and sometimes, the cat's eyes and nose become so crusted with mucus that the cat can't see or smell.

Cats with herpesvirus or conjunctivitis strongly benefit from warm compresses to clean up discharge and crust around their eyes and to remove mucus from their face and nostrils.

Physiological and psychological stress are huge factors in the prevalence of herpes outbreaks. This upper respiratory infection is common in overcrowded and poorly sanitized or ventilated animal shelters, kennels, catteries, and hoarding situations.

Since stress is such a factor in the prevalence of herpes outbreaks, supportive care, good nutrition, hygiene, environmental enrichment, and medication to treat secondary infections and discomfort are important to make cats feel better and prevent outbreaks. If proper care and nutrition are provided to the cat, symptoms should improve within two weeks.

Kittens and young cats who have suffered from feline herpesvirus are susceptible to outbreaks as adults especially during times of physical or emotional stress. Since so many cats have been affected, there are many adult cats who have recurring conjunctivitis or routinely have crust in the corner of their eyes or around their eyelids, even though they otherwise seem fine. These cats can benefit from regular warm eye compresses and gentle cleaning of their eyes.

FELINE CALICIVIRUS

The calicivirus is a common upper respiratory infection in kittens and cats with compromised immune systems. Kittens are especially vulnerable since their immune systems have not fully developed. Calicivirus causes stomatitis (mouth and tongue ulcerations) and gingivitis, as well as congested breathing and pneumonia. Some strains cause temporary arthritis in kittens and cats which causes them to limp. The virus is resistant to disinfectants so it can stay on surfaces which can make it harder to eradicate when there is an outbreak.

It is most commonly seen in high stress and overcrowded environments and unsanitary or poorly ventilated conditions such as animal shelters, kennels, catteries or breeding facilities, and hoarding situations.

Treatment is providing the cat with a clean, quiet environment, supportive care such as cleaning mucus and removing sticky secretions from her eyes and nostrils, pain killers and antibiotics to treat secondary infections and discomfort, and behavioral and environmental enrichment.

8

SOCIETAL CONCERNS

FOOD FOR THOUGHT

One primary thing I have seen over the years is an obsession with cats on diets and societal pressure, when often these cats are not, even slightly, overweight. I am contacted regularly by cat parents who are having difficulties with their cats due to their cats being put on restricted meals. Often these cat parents are in tears or at their 'wit's end' with their cats, and are ready to rehome them due to behaviors that result from their cats being hungry or ravenous.

The vast majority of cats self-regulate. When cats are free-fed, their moods change. They are happier, better-behaved, and more content. Play aggression, night wailing and excessive vocalization, inter-cat aggression, and even petting aggression diminish or disappear altogether. It may seem a 'controversial' stance, but I am not wavering on my recommendation to free-feed cats because I've seen far too many hungry and stressed cats behaviorally blossom with this simple dietary change. If you must restrict your cat's caloric intact, it is better to switch the *types* of food you feed your cat, such as switching to a lower-calorie food or adding more wet food. You might consider changing feeding strategies, such as scattering food on a tray so it takes longer for your cat to eat, or smearing wet food on a lickmat, in addition to increasing your cat's activity levels by adding vertical platforms and playing with your cat.

Cats do not handle stress well and stress makes cats sick. I've seen many happy fat cats live to a ripe old age. I have yet to see a happy cat on a diet. If your cat is begging you for food or raiding the bread basket, your cat needs to be free-fed. The bottom line: is it better to bring your cat to a shelter or get rid of your cat due to behaviors that are caused by meal restriction, or is it better for your cat to stay in the home, being better-behaved and satisfied? The trend I have seen over the last ten years is enforced diets creating far too many stressed-out and hungry food-deprived cats.

CATS IN SHELTERS & REHOMING

Sadly, there are so many animals caught in the rescue or shelter system, and even more who are strays. The bottom line is that there are millions of homeless cats and kittens. It is not atypical for me to go into a home where a cat has had multiple homes or prior owners. Cats adopted out by one shelter often end up at another, regardless of personality.

DOI: 10.1201/9781003351801-8

People have the misconception that a cat bought from a breeder or a 'purebred' is inherently better, whether better-adjusted, better-behaved, or healthier than a mixed-breed cat or one adopted from an animal shelter or rescue. However, the opposite is true. Mixed-breeds tend to be genetically healthier than purebreds.

Purebred cats became popular among the upper and middle class with the advent of cat shows in Victorian England. Cat breeding became a hobby. Cats were selected for physical appearances, not for behavior or health. Over time, cat show judges selected and awarded points for more exaggerated features between breeds. The flat, smushed-in faces of Persians and the narrow, skeletal face of Siamese cats are two examples.

Inbreeding is the norm in purebred lines, even among responsible breeders. For this reason, genetic ailments and diseases can be worsened or magnified. Nearly all purebred cats are suspectable to certain diseases and prone to medical problems. For instance, Persian cats suffer from serious ocular, respiratory, and pharyngeal problems. Their furry coats make it difficult for them to groom themselves and they overheat easily. Nature would not select traits that often cat fanciers desire for cat shows. There are genetic and biological reasons for this.

Most behavioral characteristics or personality traits people attribute to a breed are normal cat behavior. Social behavior stems from individual development and exposure to and experiences with different people, environments, and animals. Cats have not been bred for personality or to do a particular job. Being 'friendly,' good with children, cuddly, affectionate, active, liking water, and being vocal are *not* breed-specific behaviors.

Shelter animals are often stigmatized or perceived as 'less than,' even though they end up at shelters for no fault of their own. People purposely breed cats or let their cats breed and then inundate local animal shelters with unwanted kittens. Contrary to what many believe, purebred cats routinely end up at animal shelters too, especially Siamese, Persians, Bengals, Maine Coons, and Himalayans, because these breeds are so popular. Many people want a kitten, but not many people want to take responsibility for that cat or to make a long-term commitment to that animal for 15–20 years, which is the average lifespan of a cat. Although it's fun to have a kitten, within six to eight months, she will look like an adult cat. People often don't want to adopt five-, six-, or seven-year-old cats, even though these cats are still young.

Shelters don't address the problems of lack of spay–neuter laws or irresponsible pet ownership. Cat rescue organizations are often poorly funded and made up of volunteers who can only do so much at one time. Due to the sheer number of animals dropped off at one large animal shelter, it is nearly impossible to give any cat individual attention. A large city shelter can receive hundreds of kittens per day, especially during 'kitten season' which generally runs from March to September.

Because animal shelters are often inundated and overwhelmed with cats, newborn kittens or neonates are often killed upon intake. The reason for this is that newborn kittens do not thrive very well without their mother and they are extremely vulnerable to illness and disease. There is not enough time, resources, volunteers, or fosters to bottle-feed and take care of neonatal kittens.

Additionally, most animal shelters and many rescues have little to no screening for adopters and limited follow-up. This is especially true for large adopt-a-thon events and 'empty the shelter' days or for free-giveaways. Most city and county run animal shelters don't require background checks, veterinary references, or home visits for adopters.

Because of the sheer number of animals who end up at shelters on any given day, there is usually a limited or maximum length of stay before they are killed. For many animal control facilities and city shelters, this may be only 3–7 days.

Cats in shelters and stressful environments don't present well. Cats who would otherwise be social and friendly can end up hiding in a corner of a cage or meowing frantically to get out. When trapped or cornered, cats feel threatened. Being confined to a cage with no place to hide while strangers approach or try to handle them can make cats who would otherwise be gentle, hiss, lunge, and act aggressively. It is nearly impossible to tell whether a cat is 'feral' or aggressive by looking at him, evaluating his behavior in a stressful environment, or during or after physical or emotional trauma.

Stress impairs the immune system and cats who are stressed are susceptible to illness. Cats in catteries (the feline equivalent of puppy mills) and cats in loud, crowded environments are far more likely to develop upper respiratory infections and cystitis. Stress has been found to increase the likelihood of herpes outbreaks and a cat's susceptibility to feline leukemia, as well as feline infectious peritonitis.

It is vital for animal shelters to enhance the well-being of cats in their care through changes or modifications to shelter design, implementation of enrichment programs and humane handling protocols, and by providing pain relief and veterinary care. The problem is that there is no one person responsible for a homeless animal's care and no accountability if humane treatment or care aren't provided.

Animal shelters need to have compassionate, proficient leaders and those in higher positions who prioritize animal welfare and enrichment programs, better education and screening of adopters, and mandatory training for shelter staff, employees, and volunteers on the humane, gentle handling of cats.

When there is a surplus of cats and so many cats and kittens in need of homes, there is no reason to shop for or buy a cat. If you want to have an animal companion and to provide a good life to a cat, rescue or adopt one from an animal shelter or rescue organization. Advocate for better animal protection laws, volunteer, or foster. We are the only ones who can advocate for cats and how we treat them.

TNR (TRAP-NEUTER-RELEASE)

Feral colonies, as they are often called, are colonies of stray cats. Some colonies are in rural areas, but most colonies are in or near cities and towns. They are usually located in industrial areas, around vacant lots, near factories and dumpsters, and behind shopping malls and restaurants.

Cats who are lost and abandoned frequently end up at these colonies. Often people dump their cats in these places, too. Feral colonies are comprised of both feral and previously owned cats. These cats generally rely on food, scraps, and water left by people which is why they are often found in populated areas.

Intact male cats spray and fight with each other. Females go into heat and wail, and fighting can escalate between both sexes. Intact males have a wider roaming range than females and travel further distances, so they can impregnate females in many areas. Female cats tend to congregate together and stay in close proximity.

The traditional solution to unwanted and homeless cats has been to kill them by poisoning, trapping, or shooting them, often called 'euthanasia' or 'culling.' This is the approach and course of action routinely taken by state Fish and Wildlife Departments and local officials. Culling or exterminating cats is not euthanasia. Euthanasia means 'mercy' killing and is used when an animal is suffering or in pain. Killing animals because someone does not want them there or like them does not constitute euthanasia.

However, an alternative to killing cats is to trap, surgically sterilize, and then release them back into the area they were originally trapped. The tip of the cat's left ear is cut or clipped which is a visual signal to tells others that the cat has been trapped and altered so they are not unnecessarily trapped and put through surgery again. On occasion, the right ear may be clipped. A cut or clipped ear does *not* mean the cat is feral, unfriendly, or poorly socialized (Figure 8.1).

Figure 8.1 The left ear is usually clipped on cats who have been trapped, surgically sterilized, and released. © Lily Banse.

THE BENEFITS OF TNR

Aggression between cats is significantly reduced after neutering. Fighting between cats decreases and they become friendlier to each other. Neutered males tend to remain with females, instead of wandering, so they stay part of the colony.

Trapping, sterilizing, and releasing cats let them exist while lessening population growth, thereby preventing further suffering and the birth of yet more homeless animals.

People can learn to respect and coexist with homeless cats and take responsibility for them.

Good trap-neuter-release programs should ensure that:

- All trapped cats are checked for microchips or tattoos.
- There is a safe release site for the cats with someone to follow up on their care.
- Pain killer is used for any surgeries or medical treatments.
- There is a reliable food source for the cats, once they are returned, and reliable feeders. Ideally, there should be more than one person to feed the cats since there should be backup, if any person becomes ill or can't make it to the cats that day.
- Trapping efforts are maintained because people routinely abandon cats at these colonies and new cats migrate into them.
- The cats are monitored, so it is known how many cats are in the colony and which ones are new to the group.
- Medical care or treatment is provided if any of the cats become ill or are injured.

Although these suggestions are reasonable and ethical, they can be challenging to implement, so often aren't.

CHALLENGES

Although killing and extermination usually has government support, TNR programs are primarily implemented by volunteers.

Overseeing and monitoring a colony can be overwhelming for one person, so multiple caretakers are needed. These caretakers need to be reliable. All voluntary feeders and individuals monitoring these colonies should be supported. Instead, they are often punished or stigmatized by law enforcement, legislators, and the local community.

Costs of spay–neuter surgeries and veterinary treatment for multiple cats can be astronomical and overwhelming.

Without a supply of food, safe shelter, and veterinary attention, feral and stray cats die of illness, injury, or starvation.

Human ignorance regarding cat behavior and welfare is pervasive, including the vilification of cats for hunting, instead of human culpability for bird deaths caused by window collisions, grass lawns and lawn chemicals, habitat destruction, fishing hooks, glue traps, rodenticides and pesticides, unregulated trapping, and development.

Access to reliable veterinary care and affordable sterilization services for homeless and stray cats is limited. If veterinarians won't provide veterinary treatment to unowned cats and don't provide safe or effective sterilization services, who is going to?

There is a misconception that feral cats can't be rehomed. Many people don't know how to deal with frightened or defensive cats and don't have the patience for cats who are fearful or traumatized. People want cats to be affectionate, docile, or social even upon rough handling, trapping, approach by strangers, and after invasive surgery. This is unrealistic. It can take a few weeks or even months for traumatized or frightened cats to become docile and friendly. Unfortunately, too often, these cats are labeled 'feral,' aggressive, fractious, or difficult to handle and immediately marked unadoptable. How cats behave in a veterinary setting or under stressful circumstances is not indicative of how they will behave in a home or non-veterinary environment.

There is a lack of support from people within the community, local government, and law enforcement. The harder they make it for people to help cats, the less progress is made and they become an impediment to effective solutions.

Cats continue to hunt after being neutered, so people often want 'feral' cats exterminated thinking that they harm wildlife, not realizing that most homeless cats, especially in cities and suburbs, don't hunt, but rely on people for handouts and scavenge (Figure 8.2).

There is a culture of indifference in our society to homeless animals and adopting animals from shelters as opposed to buying them online and from breeders. Unless purchasing cats from catteries and online is made socially unacceptable, cats who end up at shelters will languish or, if not adopted, be killed for space.

CRITICISMS AND CONTROVERSY

TNR is not universally supported among the veterinary profession. For some, it's been called 'trap, neuter, abandon (or re-abandon).'

As is human nature, it's easier to criticize but harder to find or create real solutions, especially to multifaceted problems. The criticisms, however, are valid, but challenging to address and apply to how animal shelters operate as well, even well-funded ones.

When cats are sterilized through low-cost spay–neutering services, painkillers are seldom used due to cost. There is rarely follow-up care or recovery time for cats after being subjected to surgery.

Figure 8.2 Homeless, stray, and feral cats in cities, suburbs, and towns generally don't hunt, but mostly scavenge and rely on people for food. © Olha Novytska.

After surgery with general anesthesia, females cats require at least a few days to recover before they should be returned. When owned cats are altered, the veterinary recommendation is some form of collar or monitoring the cat for 7–10 days to prevent her from opening up or licking any sutures. This is not an option for most cats in trap-neuter-release programs. Likewise, there is no follow-up care or re-checks for the cat to ensure there were no complications.

Animal shelters and trap-neuter-release programs routinely separate mothers from kittens too early. Four to six weeks is too young to separate a mother from her kittens and early-weaned kittens often have behavioral problems which can later end up with them being dumped or brought to shelters. If kittens younger than three weeks are provided care, they go to fosters and are bottle-fed. However, many kittens don't survive and the mother, who may be a kitten herself, is released and has to fend for herself.

Releasing cats back to an unsafe environment is irresponsible, especially if the cat was in poor condition when found. If a cat is suffering from illness or injury, if the environment is unsafe, or if a location is going to be torn down or developed, he should be relocated or rehomed, not returned. There is the argument that euthanasia is better than returning a cat to a poor existence with no follow-up care, kindness, or care-givers.

Health disorders can't be treated, injuries can't be healed, and illnesses can't be managed. Safety can't be insured, and food is not predictable or constant.

Cats when trapped are terrified. If they are left in traps, they can seriously injure themselves due to panic. For TNR, if cats have no safe place to be released while they wait to be neutered, they can remain in a trap for up to a few days.

Cats are not safe outdoors even in managed colonies, especially from violent or sadistic individuals. Research has long shown that those who perpetrate violent crimes against people select cats for cruelty. Because animal protection laws are few and there are no real consequences for animal abusers, these people can abuse cats without getting caught and with no repercussions.

Most TNR programs and animal shelters test cats for FIV and FeLV upon intake or after trapping. Cats who test positively are usually killed. This is a mistake and is especially relevant for kittens. Kittens will test positive because of maternal antibodies they received from the mother and they can test falsely positive for up to six months. Cats who have been vaccinated against FIV and FeLV, which is now more common, test positive even though they will never have the disease or pass it to others. Instead of blanket testing, cats showing signs of illness should be cared for in the same way as any other cat.

The continued monitoring of colonies is vital and trapping and neutering has to be on-going. Within a short amount of time, a new group of cats can be born by only just one intact male and female. It can take only minutes to copulate. It can take days or a few weeks to trap.

SOLUTIONS

There are no clear-cut solutions that can be implemented without support from local governments and county officials. When it comes to implementing animal protection laws and animal advocacy, the legwork and power-lifting is usually done by only a few individuals who are rarely supported or recognized. Media and local communities must be involved and supportive as well.

There should be mandatory spaying and neutering of owned cats. When there is legislation regarding sterilization of animals, it is usually vehemently fought by the American Kennel Club, breeders, and sporting clubs, pet shop owners, and individuals who profit from breeding or selling animals. Although surprising to some, the well-funded American Veterinary Medical Association usually fights against animal protection legislation, not for it.

Veterinary schools need to teach the importance of feline population management programs and low-cost spay–neutering services to veterinary students which, at this point, is rarely done.

Just as there is a requirement for licensing dogs, there can be one for registering cats. Although it may sound strange because it isn't done, the benefits can be accountability if a cat is abandoned or abused. Cats can be reunited with their owners when they are lost. Discounts or free registration can be given for cats who are vaccinated and sterilized which gives people an incentive to do so. Problems with implementing this are oversight and enforcement, as well as legislators involving themselves with enacting pet limits. This usually isn't the case for dogs, so shouldn't be made an issue for cats.

Changing the laws regarding pet ownership and rental housing requirements, as well as condo and HOA (home owner association) rules. Housing laws greatly discriminate against people with animals. Many healthy, well-cared-for animals are relinquished to shelters because property owners implement no pet or one pet policies on their rental units, or they charge extra, unnecessary fees, such as pet rent. These fees don't apply to the number of tenants in a dwelling or the number of human children a tenant might have.

Accessible and affordable sterilization services need to be made available for everyone. Veterinary fees are unaffordable for a large percentage of the population. Unfortunately, veterinary schools and the AVMA (American Veterinary Medical Association) are resistant to having upper level veterinary students provide pro-bono veterinary care or low-cost spay–neutering services. Until veterinary medicine becomes more affordable and accessible, there will continue to be animals who are neglected and uncared for.

More powerful and meaningful animal protective legislation must be enacted. Animal welfare legislation, if passed, is usually watered down with loopholes making the laws unenforceable or open to interpretation. Large not-for-profit animal protection organizations are frequently responsible because it looks good on paper to get a bill passed (they get more donors), but when it comes to the bill having real substance, it doesn't. A poor animal protection law that legalizes or legitimizes animal cruelty can be worse than no law at all, because once a law is enacted, it becomes harder to change. If there are no severe consequences for animal cruelty, neglect, and abuse, then it gives the public permission to physically abuse, neglect, and abandon animals.

Society needs to make it the new normal to adopt cats from rescue organizations and animal shelters, and socially unacceptable to buy cats online and from breeders. This should be encouraged by public figures, veterinary professionals, local governments, on social media, and in other media outlets.

My personal take is that much of the responsibility lies on the shoulders of large multi-million dollar 'not-for-profit' animal protection organizations. These organizations are well-funded and have the man power to make effective change, but often do little legislatively. While the NRA, hunting, and breeding lobbies are in the courts, meeting with legislators and enacting or dismantling legislation, large animal protection organizations throw galas, compete for donors, and mingle with celebrities. They operate more like 'for-profit' corporations with big CEO salaries and well-funded marketing campaigns.

Larger and better-funded organizations should connect people at local levels to be politically active and empower individuals, local activists, and rescue groups so that they can network and support each other. Instead, very large and well-funded organizations tend to compete over donors and memberships, spend huge sums on PR and marketing, and often take credit for smaller organizations' and other animal advocates' work and accomplishments. Unless these organizations are held accountable and local groups and individuals support each other, little may change.

GLOSSARY/BEHAVIORAL TERMINOLOGY

FELINE LEARNING & BEHAVIOR MODIFICATION PRINCIPLES

Behavior Modification: Using the principles of how animals learn to change behavior. More technically, changing behavior through classical and operant conditioning to replace unwanted behaviors with preferred ones.

Stimulus: Something (internal or external) that causes a physiological (physical, behavioral, or emotional) response or change in an organism.

Stressor: A stimulus or something that causes stress, such as a loud voice or icky smell.

Stress: Psychological reaction to a trigger that causes the cat to become upset, anxious, or unhappy.

Desensitization: Exposing the animal to a stimulus at a level that causes no negative reaction. Incremental exposure to something at the cat's comfort level. The cat experiences little to no fear or anxiety upon exposure. Only when the cat is emotionally and physically comfortable, do you increase exposure to the stimulus.

Classical Conditioning: Process of learning by association. Many fear responses are due to negative associations learned through classical conditioning. Classical conditioning happens continually, and we are often unaware of it.

> *Example:* Every time you open a can of cat food, it makes a specific sound, and every time you open a can of cat food, you feed your cat. Eventually, cats associate a can opening with food and being fed.

> *Example:* The carrier comes out of the closet and the cat is put into it, taken for a frightening car ride, and roughly handled at the veterinary clinic. Upon return, the cat is released from the carrier and the carrier is put back into the closet. Whenever the carrier comes out, the cat is forced into it again and goes for a car ride to the vet. The cat develops a negative association with the carrier. The moment he sees or hears it, he runs and hides. He now associates the carrier with car rides and rough handling at the veterinary hospital.

Counter-conditioning: Classically conditioning a different or opposite association.

> *Example:* A cat was previously brushed with a metal-bristled brush which caused him discomfort so he associates the brush with pain. He swats at the brush or runs away upon seeing it. If we desensitize him to the brush, we can start with a new brush with soft bristles. We can brush him gently once or twice in the areas he likes and when he is affectionate on the bed. When we brush him and afterwards, we give him a pile of treats. We continue to pair the brush and being brushed with cuddle time and treats on the bed. Now, instead of running away upon seeing the brush, he hops onto the bed in anticipation of treats and cuddle time. We classically counter-conditioned him to the brush since he now has a positive association upon seeing it.

Conditioned Response: A response in an animal that is classically conditioned.
Conditioned Stimulus: A stimulus that causes a conditioned response.

Desensitization and counter-conditioning are often used in conjunction with each other to resolve behavioral problems.

Operant Conditioning: Learning by trial and error. Animals repeat behaviors that are reinforced or rewarded (There are self-reinforcing behaviors, too) and avoid or stop behaviors that are not reinforcing or rewarded.

There are four categories of operant conditioning: Positive reinforcement, negative reinforcement, positive punishment, and negative punishment. Techniques can fall into more than one category depending on how the behavior or technique is described or explained. So, don't worry too much about labels or technical terms.

Reinforcement & Punishment

Reinforcement: Something that increases the likelihood of a behavior.

Punishment: Something that decreases the likelihood of a behavior.

Positive Reinforcement: The repetition or frequency of a behavior increases when it is rewarded (A reward does not have to be food. It is whatever the animal is wanting or needing at the time or something the animal finds desirable).

Negative Reinforcement: Stopping or removing something unpleasant or aversive, so the likelihood of a behavior will increase. In other words, the aversive or painful stimulus is removed when a correct or wanted behavior is performed. Learning this way is hard, although, unfortunately, it's often how humans teach animals.

Positive Punishment: To apply something aversive or unpleasant to decrease the likelihood or repetition of a behavior.

Negative Punishment: To remove something desirable or pleasant for the animal (something she wants) to decrease the likelihood or frequency of a behavior.

Positive reinforcement and negative punishment go hand in hand and are often used in conjunction with each other. Likewise, positive punishment is often used with negative reinforcement.

Examples:
Positive Reinforcement: The cat paws at our leg for attention. We pick up the cat and give her a kiss. Her behavior becomes more frequent and common. By giving the cat what she wants or desires, we positively reinforced the behavior.

Negative Punishment: When the cat paws at your leg, you remove your attention or look away from her – each and every time. She stops pawing your leg to get attention. (She may find something else to do instead!)

Positive Punishment: When examining the cat, she starts to squirm. We scruff her neck and pin her to the table so she stops. After a minute, she stops struggling.

Negative Reinforcement: We stop scruffing or pinning the cat and let go once she remains still or is quiet. In other words, the aversive or punishment is removed or stops when the cat does what we want.

Don't worry about labels! You'll have all the knowledge to resolve your cat's behavior problems throughout this book without having to know technical definitions.

INDEX